GEOPHYSICAL METHODS IN GEOLOGY

SERIES

Methods in Geochemistry and Geophysics

1. A.S. RITCHIE

CHROMATOGRAPHY IN GEOLOGY

2. R. BOWEN

PALEOTEMPERATURE ANALYSIS

3. D.S. PARASNIS

MINING GEOPHYSICS

4. I. ADLER

X-RAY EMISSION SPECTROGRAPHY IN GEOLOGY

5. THE LORD ENERGLYN AND L. BREALY

ANALYTICAL GEOCHEMISTRY

6. A.J. EASTON

CHEMICAL ANALYSIS OF SILICATE ROCKS

7. E.E. ANGINO AND G.K. BILLINGS

ATOMIC ABSORPTION SPECTROMETRY IN GEOLOGY

8. A. VOLBORTH

ELEMENTAL ANALYSIS IN GEOCHEMISTRY
A: MAJOR ELEMENTS

9. P.K. BHATTACHARYA AND H.P. PATRA

DIRECT CURRENT GEOELECTRIC SOUNDING

10. J.A.S. ADAMS AND P. GASPARINI

GAMMA-RAY SPECTROMETRY OF ROCKS

11. W. ERNST

GEOCHEMICAL FACIES ANALYSIS

Methods in Geochemistry and Geophysics

12

GEOPHYSICAL METHODS IN GEOLOGY

by

P.V. SHARMA

Professor of Solid Earth Physics, Institute of Geophysics
University of Copenhagen, Copenhagen, Denmark

Elsevier · New York

NEW YORK · OXFORD

ELSEVIER SCIENTIFIC PUBLISHING COMPANY
335 JAN VAN GALENSTRAAT
P.O. Box 211, AMSTERDAM, THE NETHERLANDS

ELSEVIER NORTH-HOLLAND, INC.
52 VANDERBILT AVENUE
NEW YORK, NEW YORK 10017

Library of Congress Cataloging in Publication Data

Sharma, P   Vallabh.
  Geophysical methods in geology.

   (Methods in geochemistry and geophysics ; 12)
   Bibliography: p.
   Includes index.
   1. Geophysics. 2. Prospecting—Geophysical
methods. 3. Geology—Methodology. I. Title.
QE501.3.S48   551'.028   74-77586
ISBN 0-444-41235-2
ISBN 0-444-41383-9 pbk.

WITH 193 ILLUSTRATIONS AND 15 TABLES

Second Printing, 1978

Manufactured in the United States of America

"Earth science is a particularly alluring field for premature attempts at the explanation of imperfectly understood data."

R.H. DANA

Preface

This book developed from a course called "Geophysical Methods in Geology" that I organized at the University of Copenhagen in 1967. The course was aimed at providing a concise but fairly comprehensive introduction to the methods of geophysics within which the undergraduate students of geology and geophysics could locate their specific interests for further specialization.

For the earth science student who is interested in the broader aspects of solid earth geophysics, including the various ways in which it can be of help in structural, correlation and geochronological problems, no suitable text is available. Several introductory texts on "applied" (exploration) geophysics exist (e.g., Dobrin, 1960; Griffiths and King, 1965; Parasnis, 1972, 1973); in which the major emphasis is on problems relating to oil and ore prospecting or to investigations of shallow structures for civil engineering needs. In "pure" geophysics, on the other hand, all the modern texts in English (e.g., Stacey, 1969; Jeffreys, 1970; Bott, 1971; Garland, 1971) are at a fairly advanced level and in these texts the relative emphasis is on the physics of the earth's interior. The realization of this lack of an introductory text on geophysical methods with a broad range of applications in geology led to the present work.

In aiming the book at essentially two groups of students, those from both geology and physics, I am well aware of the pitfalls. The challenge is to steer a middle course as best one can between them. It is assumed that the student using the text will be familiar with the terminology of both disciplines. Two semesters (one year) of geology and two years of college physics should be a basic preparation for understanding the material presented here. In view of the general observation that the average geology student is unfamiliar with higher mathematics, its use has been kept at a minimum. Except in a few cases, no mathematical formulas are derived and the reader is asked to accept some statements on trust. I believe this is unlikely to seriously impair the value of the book, for the interested reader can always look for the derivations and details in the references cited. The underlying philosophy has been that, in understanding the earth's phenomena, physical ideas matter more than mathematical derivations.

No claim is made that the book is comprehensive in the sense of including important aspects of every geophysical method. The field is so wide that a selection of topics, with some bias resulting from personal interest, is inevitable. The relative emphasis is on those topics whose importance has considerably increased in the last decade, which has been proclaimed by many as the decade of "revolution" in the earth sciences. These topics include geophysical studies of coastal margins, ocean ridges and trenches, palaeomagnetism, geomagnetic reversals, heat flow, seismotectonics, and plate tectonics. Some problems of exploration geophysics are also discussed as there are several points of contact between academic studies and prospecting surveys. Among the principal omissions are the electromagnetic methods, seismographs and the arrays, and some current topics like prediction and artificial release of earthquakes. Perhaps this is to be regretted, but, with their inclusion, the contents could easily have got out of hand, and, with some justification, the book was to be kept to a tolerable size.

The scope of the students' interest in geophysical methods and their applications is by no means uniform. With this in mind I have tried to arrange the subject matter so that the reader may find most of what he is particularly interested in without having to study the whole of the book. Thus, each of the chapters on seismic, gravity, magnetic, resistivity, radioactivity, and geothermal methods can be read independently. The last chapter dealing with global tectonic theories may, however, not be easy to follow without reading the preceding chapters.

In this book I have generally adopted the Système Internationale (SI) units. (The SI system is an extension of the rationalized MKSA system.) However, in a few instances, I have departed from the exclusive use of the SI system in using some of the widely accepted "working units", e.g., g/cm^3 for density, *mgal* in gravity studies, the *Richter magnitude* scale for earthquakes, and *cm/year* for the ocean-floor spreading rates. Also, to avoid redrawing a number of figures taken from other sources, old units have been retained in some figures and a note of conversion to SI-units is added either in the text or in the figure caption. A brief conversion table of geophysical units is included as an appendix.

An attempt has been made to keep the book broadly up to date in the selection of material and references from the increasingly voluminous literature. For the selection of recent references, it must be admitted that these have been chosen with the English-speaking reader in mind.

During the preparation of the text I received help from many friends and colleagues in numerous ways. I wish to thank all of them and in particular the following who critically read parts of the manuscript at various stages: R.N. Athavale, M. Båth, H. Berckhemer, I. Lehmann, D.S. Parasnis, and T.C.R. Pulvertaft.

The kind cooperation of many authors and publishers who extended permission to use their data and figures is greatly appreciated.

Copenhagen   P.V. SHARMA
1974

Copyright Acknowledgements

I am grateful to the authors and the following publishers for permission to reproduce their diagrams as detailed below. Credit to the authors is also given in the figure captions.

The American Geophysical Union, Washington, D.C., for Figs. 14, 44, and 46 (© 1969 A.G.U., Geophysical Monograph, No. 13, pp. 263, 204, and 221 respectively).

McGraw-Hill Book Company, New York, N.Y., for Fig. 32 (from Grant and West, 1965, p. 147), Figs. 40 and 58 (from Nettleton, 1940, figs. 125 and 27 respectively) and Fig. 157 (from Jacobs et al., 1959, fig. 88).

Pergamon Press, Oxford, for Figs. 49 and 72 (from Garland, 1965, figs. 3.3 and 8.1 respectively) and Fig. 52 (from Griffiths and King, 1965, fig. 6.6).

Academic Press, New York, N.Y., for Fig. 63 (from Gutenberg, 1959, p. 194) and Fig. 178 (from Runcorn, 1962, fig. 20).

John Wiley and Sons, Inc., New York, N.Y., for Figs. 89 and 131 (from Irving, 1964, figs. 2.3 and 6.28 respectively).

Thomas Nelson, London, for Fig. 122 (from Holmes, 1965, fig. 865).

Contents

CHAPTER 1

The Scope of Geophysics and Methods

By definition, geophysics is the scientific study of the earth using methods of physics. Along with geology and geography it occupies an important position in earth sciences.

Strictly speaking, the subject covers the physics of the whole earth, from its deepest interior to the extreme fringe of its atmosphere, and so involves many disciplines. However, in current practice it is often used in a more restricted sense to denote the physics applied to the study of the "solid earth" (excluding the hydrosphere and atmosphere), and it is in this sense that the word geophysics will be used in this book.

The domain of geophysics, even in its restricted meaning, involves many fields of study (Fig.1). These are:

Geodesy and gravimetry—dealing with the earth's shape and gravitational field.

Seismology—dealing with earthquakes and other ground vibrations (e.g. that caused by chemical or nuclear explosions).

Geomagnetism and geoelectricity—dealing with the earth's magnetic and electric phenomena.

Geothermometry—dealing with the heat flow and temperature distribution in the earth.

Tectonophysics—dealing with the physical aspects of regional and global tectonics.

Geocosmogony—dealing with the origin of the earth.

Geochronology—the dating of events in the earth's history.

In addition there are some fields of study which, although generally recognized as belonging to geophysics, are more closely linked with geology. These are specialized disciplines

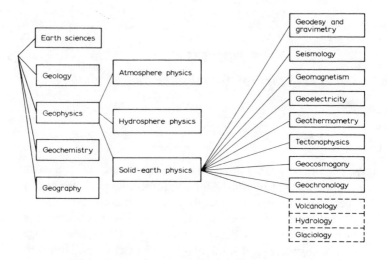

Fig. 1. Schematic classification of the earth sciences into component disciplines. Important divisions of geophysics and subdivisions of the "solid-earth physics" are also shown.

such as *volcanology, hydrology* and *glaciology* whose fields of interest are evident from their names.

It is now recognized that any attempt to rigidly define the boundaries of the component disciplines of geophysics would be futile. This is so because, to a large degree, they are interdependent, and work as complementary tools in the hands of an earth scientist. For example, geothermometry, which deals with the heat of the earth, is related not only to volcanology, seismology and radioactivity but also to geocosmogony and tectonophysics. To mention another example, the subject of tectonophysics is so closely linked with seismology, geothermometry, volcanology, gravity, and rock magnetism, that it is impossible to study it as a completely separate subject.

As for geophysics itself, it is almost impossible to study it as an isolated discipline. Geophysics developed from the disciplines of physics and geology and has no sharp boundaries which distinguish it from either. In its methods, geophysics is a branch of experimental physics, but in studying the various complexities of the earth it shares with geology the common objective of understanding the planet we inhabit. Geophysicists have been more concerned with actual measurements of earth phenomena—seismic energy travel times, gravity and magnetic field measurements, and so forth. These have been considered to be essentially *quantitative* and *tangible*. Geologists are more concerned with *qualitative* and *descriptive* studies, although paradoxically every field geologist begins with quantitative measurements such as the thickness and the dip of a specific surface formation.

As the tools used by geologists become more and more refined in precision, geologists progressively use a greater variety and quantity of tangible measurements. As geophysicists are required to seek progressively deeper prospects (or phenomena) in the earth, some of their measurements become less easy to interpret. As Birch (1952, p.234) has remarked: "Unwary readers should take warning that ordinary language undergoes modification to a high-pressure form* when applied to the interior of the Earth...."

Therefore the line of distinction between what is quantitative and what is qualitative tends to fade, and with it the traditional distinction between geophysicists and geologists. Today, their formerly discrete endeavours are becoming better coordinated. This trend is quite apparent in exploration geophysics where the majority of petroleum geologists spend most of their time extracting quantitative information from geophysical data such as seismic record sections, electrical and a family of other well logs. Likewise, geophysicists who have been largely concerned with measurements of physical phenomena, are nowadays incorporating more geology in order to increase the reliability of the conclusions.

*A few examples of high-pressure equivalents of ordinary meanings (in brackets) are: certain (dubious); undoubtedly (perhaps); positive proof (vague suggestion).

Clearly, the recent revolution in earth sciences, relating to the new developments in the global tectonic theory (such as "sea-floor spreading" and "plate tectonics"), is due mostly to geophysicists and geologists with two characteristics in common: a broad training in physics and geology, and an interest in worldwide problems rather than small areas.

At the university level, however, the distinction between geology and geophysics persists, and radical changes cannot be anticipated very soon. Universities will continue to produce graduate geologists and/or geophysicists instead of "earth scientists" or "explorationists" for a considerable time. Nevertheless it is gratifying to learn that at most universities a growing demand for introducing geophysics in the geological curricula is being felt. In a way this book attempts to meet this demand.

Every earth scientist, especially the geologist, should be familiar with the various methods of geophysics which can help him understand the earth. At the same time he must be aware of the limitations of the geophysical methods. It must be emphasized that in studying the earth's hidden features and phenomena, most problems are of an "inverse" type, i.e., of deducing the "source" from the observed "effect". The measured physical effect (e.g. surface variations in gravity, magnetic or electric field) in general cannot be interpreted in terms of a unique source (or phenomenon) occurring at a particular depth inside the earth. This is because a variety of sources with varying parameters at different depths can theoretically produce the same effect. A combination of several geophysical methods often yields more information that can help reduce the ambiguity by narrowing down the range of possible solutions. Also, a geological intuition combined with logic may help to further reduce the number of likely possibilities.

The physical properties of rocks that are most commonly utilized in geophysical investigations are the elasticity, density, magnetization, electric conductivity, radioactivity and the thermal conductivity. These properties have been used to devise methods for the study of the earth. It is convenient to classify these methods under six major headings:

> Seismic methods
> Gravity methods
> Magnetic methods
> Electrical methods
> Radiometric methods
> Geothermal methods

All the major geophysical methods were originally developed to investigate the earth's large-scale features. For example, early gravity measurements (with pendulums) were used to determine earth's precise shape; seismic travel-times of earthquake waves were used to unravel the structure of the earth's interior; magnetic measurements were used to record the geomagnetic elements on a global scale. As soon as it was realized that these methods could also be employed for oil and ore exploration, advances in instrumentation and techniques followed and the science of "applied" geophysics started developing rapidly. Pure and applied science by their very nature are, however, so closely linked that always when one advances it aids the progress of the other discipline. For instance, many of the techniques developed to explore for oil and minerals have been advantageously used in academic studies relating to the structure of the earth's crust and its interior. The special applications of geophysics to exploration problems are covered in several excellent texts (e.g. Parasnis, 1973; Dobrin, 1960; Gurwitsch, 1970). This book will deal with the broader aspects of geophysics, particularly the various ways in which it can help a geologist to understand the earth.

In the chapters that follow, the different geophysical methods are outlined and some of their applications to various geological problems, including those relating to global tectonics, are described.

CHAPTER 2

Seismic Methods

INTRODUCTION

The most highly developed single branch of geophysics is seismology. In fact seismology began as a science of earthquakes (*seismos* in Greek means shock) to study the causes and effects of the most calamitous natural phenomenon of the earth. However, currently this field of study covers various types of earth movements, from the large earthquake waves to the minute omnipresent seismic pulsations. The frequency range also encompasses a wide spectrum, covering high frequency vibrations ($\geqslant$100 c/s) to very slow movements with periods (the time taken to complete one cycle) of an hour or more. Thus, in seismology, we actually deal with a wide spectrum of seismic movements. Such a seismic spectrum, indicating various types of earth movements, is outlined in Fig.2.

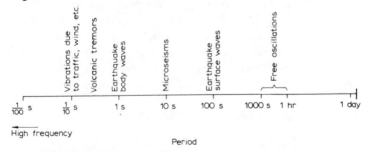

Fig.2. Seismic spectrum. The range of periods for various types of ground movements are shown in a logarithmic scale.

Speaking generally seismic methods may be classified into two major divisions, depending on the energy source of the seismic waves. One, in which the natural shock waves from earthquakes are studied to infer physical properties and structure of the earth's interior, is called *earthquake seismology*. The other, in which the seismic waves are generated by artificial explosions at selected sites to obtain information about regional or local structures, is called *explosion seismology*. This classification is somewhat artificial because the development of the observational techniques of both methods has widened their domains of applications with considerable overlapping. For instance, precisely controlled nuclear explosions have been invaluable in extracting finer details about the internal structure of the earth, whereas earthquake surface waves have contributed a lot to the study of crustal structure. In the same way the development of instrumental techniques for one method has led to the development of parallel techniques for use in the other method. Great advancement in techniques of explosion seismology is due mostly to its extensive use as a tool in oil exploration since the 1920's.

Seismology is a rapidly advancing field with a wide range of applications including earthquake seismology, seismotectonics (relationship between earthquakes and tectonic movements), and explosion seismology, which includes seismic prospecting.

This chapter aims to describe the principles and practices of various seismic methods with special emphasis on studies relating to the earth's layered structure, regional seismotectonics and seismic prospecting. The discussion of global seismotectonics and its implications for the modern concept of "plate tectonics" is deferred to a later chapter.

FUNDAMENTALS OF SEISMIC WAVE PROPAGATION

Seismic methods depend basically on the propagation of waves in elastic media. To understand the behaviour of seismic waves which travel in rock media, it is necessary first to define the quantities that describe the elastic properties of a medium. In considering the elastic properties of rocks, we assume that a rock body is homogeneous and isotropic (Fig.3), as otherwise the seismic wave propagation becomes much too complicated. In practice, this assumption simplifies the interpretation of the measured effects (time anomalies) in terms of departures from assumed uniform conditions, which occur inside the earth.

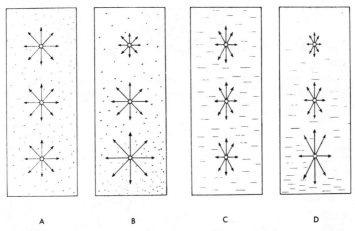

A   B   C   D

Fig.3.Schematic illustration of seismic ray velocities in four rock types: A. Isotropic and homogeneous. B. Isotropic and inhomogeneous. C. Anisotropic and homogeneous. D. Anisotropic and inhomogeneous. (After Gassman and Weber, 1960).

Elastic constants

The elastic properties of substances are described by certain elastic moduli or constants which quantitatively specify the relation between different types of stresses and strains. "Stress" is a measure of the force per unit area (F/A, in N/m^2)* associated with an elastic deformation. "Strain" is a measure of the resulting deformation and expressed per unit length ($\Delta L/L$)

* N (newton) is the SI unit of force (see Appendix).

or per unit volume ($\Delta V/V$). Within the limits of elasticity, stress is proportional to strain (Hooke's law). The important elastic constants, and the interrelationships between them, are as follows. Young's modulus, E, is a measure of the stress/strain ratio in the case of a simple tension or compression and is given by the relation :

$$E = \frac{F/A}{\Delta L/L} \qquad (2.1)$$

where ΔL is the change in length caused by a longitudinal stress.

Bulk modulus, k, is a measure of the stress/strain ratio in the case of a simple hydrostatic pressure, P, that produces a change in volume. This is expressed as :

$$k = \frac{F/A}{\Delta V/V} = \frac{P}{\Delta V/V} \qquad (2.2)$$

The reciprocal of the bulk modulus is called the compressibility.

Shear modulus or rigidity, n, is a measure of the stress strain ratio in the case of a simple tangential stress (shear). The shearing strain is the resulting deformation without change of volume (as a pile of cards can be sheared without affecting the total volume of the cards) and is often measured as an angle of deformation, ϕ :

$$n = \frac{F/A}{\Delta L/L} = \frac{F/A}{\phi} \qquad (2.3)$$

Poisson's ratio, σ, is a measure of the geometrical change in the shape of an elastic body. For instance, a cylinder of length L and diameter D, when subjected to a tensile stress parallel to L, would be elongated in length by ΔL, but at the same time it will be shortened in diameter by ΔD. In the case of a compressional stress acting parallel to L, there would be a shortening in length and an increase in diameter. In either case, Poisson's ratio is expressed as :

$$\sigma = \frac{\Delta D/D}{\Delta L/L} \qquad (2.4)$$

The value of σ can never be greater than $0\cdot5$. For most of the rocks σ is approximately $0\cdot25$.

Only two of the above-mentioned four elastic constants are independent and all four can be expressed in terms of any two :

$$\begin{aligned}
k &= E/3\ (1-2\sigma) \\
n &= E/2\ (1+\sigma) \\
E &= 9kn/(3k+n) \\
\sigma &= (3k-2n)/(6k+2n)
\end{aligned} \qquad (2.5)$$

Elastic waves

When a stress is suddenly applied to an elastic body (as when it is hit with a hammer) or when the stress is suddenly released (as when a previous state of stress is altered by fracturing), the corresponding change in strain is propagated outwards as an elastic wave. There are two principal types of elastic waves :

(1) *Body waves.* An elastic medium can be subjected to two types of deformation: compression and shear. Hence all the elastic waves detected in seismology are basically "compressional/dilatational" or "shear" waves. The essential difference between the two types is that one entails a volume change without any rotation of the elastic material element, whereas the other entails a rotation without any change of volume. In compressional/dilatational or P* waves, the particles of the

---

* P for *primus* (primary) and S for *secundus* (secondary).

medium move in the direction of wave travel, involving alternating expansion and contraction of the medium, as in the case of sound waves. The velocity of P waves (also called longitudinal waves) is given by:

$$V_\mathrm{P} = \sqrt{\left(\frac{k+4n/3}{\rho}\right)} = \sqrt{\frac{(1-\sigma)\,E}{(1+\sigma)\,(1-2\sigma)\,\rho}} \qquad (2.6)$$

where ρ is the density of the medium and σ Poisson's ratio.

The equivalence of the two expressions in eq. 2.6 follows from the interrelationships between E, n, k and σ given by the set of eqs. 2.5.

In shear or S waves, the motion of the particles of the medium is perpendicular (transverse) to the direction of wave travel (like waves on a vibrating string). Only rigid (solid) materials can transmit shear waves. Their velocity is given by :

$$V_\mathrm{S} = \sqrt{\frac{n}{\rho}} = \sqrt{\frac{E}{\rho}\,\frac{1}{2(1+\sigma)}} \qquad (2.7)$$

It follows from eqs. 2.6 and 2.7 that $V_\mathrm{P} > V_\mathrm{S}$. For most of the rocks, with $\sigma \sim 0.25$, $V_\mathrm{P} \sim 1.7\,V_\mathrm{S}$. Since the density ρ does not usually vary by as much as a factor of 2 in rocks, and since σ is usually approximately around 0.25, it also follows from eqs. 2.6 and 2.7 that the elastic parameter E is the most important variable controlling the velocity of seismic waves in rocks.

Fig.4 shows the displacement of medium particles associated with P and S waves. In seismology, both P and S waves are referred to as "body waves".

(2) *Surface waves.* In addition to the body waves which travel across an elastic medium, there are waves which travel only along the free surface of an elastic solid. There are two types of surface waves in solids. In Rayleigh waves (see Fig.5) the particle motion is more or less a combination of longitu-

Longitudinal or P-waves

⌐One wave length ⌐

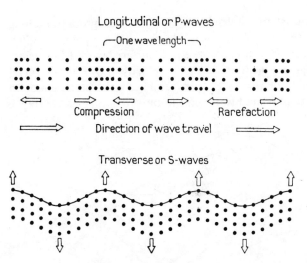

Compression   Rarefaction

Direction of wave travel

Transverse or S-waves

Fig.4. Particle movements in longitudinal and transverse seismic waves. (Courtesy of Harper and Row, Publishers.)

dinal and transverse vibration, giving rise to an elliptical retrograde motion in the vertical plane along the direction of wave travel. The velocity of Rayleigh waves is about 0·9 V_s.

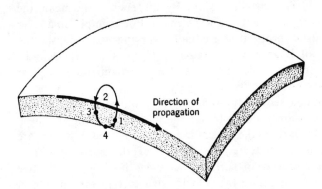

Fig.5. Schematic illustration showing the elliptical, retrograde particle motion in a vertical plane that occurs with Rayleigh wave propagation.

In contrast to Rayleigh waves which can propagate along the surface of a uniform solid, Love waves are possible only if the material is non-uniform, e.g. a surface layer of low velocity overlying a medium of higher velocity. Love waves travel horizontally in a surface layer (see Fig.6), the particle motion being horizontal and transverse to the direction of wave travel.

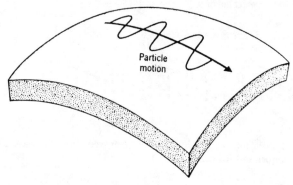

Fig.6.The particle motion associated with Love waves. The motion is essentially horizontal and transverse to the direction of wave travel.

An important characteristic of surface waves (both Rayleigh and Love) is their dispersion. By dispersion, we mean the dependence of velocity on wavelength. For example, in a layer with elastic properties gradually varying with depth, the velocity of Love waves is equal to V_s for short wavelengths in the upper part of the layer and it approximates to V_s for long wavelengths in the lower part.

In applied seismology (mainly for geological mapping and prospecting operations) by far the most important are the P waves. In earthquake seismology, both P and S waves are important in studying the interior of the earth. Recently dispersion studies of surface waves has become an effective tool for investigating the velocity structure in the outer part of the earth that complements the information available from body waves.

Reflection and refraction of waves

In an homogeneous medium, P and S waves extend uniformly in all directions from the source of disturbance; advancing wavefronts are spherical surfaces centred at the source and perpendicular to the direction of propagation. The energy of a wave declines rapidly with the square of the distance, whereas the amplitude reduces directly in proportion to the travel distance.

When a disturbance impinges on the boundary of a second medium with a different elastic velocity, energy is partly reflected and partly transmitted (refracted) into the second medium. The basic concepts governing the reflection and refraction of seismic waves are the same as in geometrical optics and, at large distances from the source, the approximation of wave paths by rays is equally valid. Therefore, both Huygen's and Fermat's principle are applicable to seismic waves. Despite this similarity, the actual processes of reflection and refraction of seismic waves are a little more complicated than those of light waves, since in general any P (or S) wave striking a boundary will generate two reflected (P and S) and two refracted (P and S) waves. Using the notation of Fig.7, the laws of reflection and refraction are given by :

$$\frac{\sin i_P}{V_{P_1}} = \frac{\sin R_P}{V_{P_1}} = \frac{\sin R_S}{V_{S_1}} = \frac{\sin r_P}{V_{P_2}} = \frac{\sin r_S}{V_{S_2}} \qquad (2.8)$$

The equality of the angles of incidence and reflection (e.g. $i_P = R_P$) holds only if the incident and reflected waves are of the same type.

If we consider for the present only P waves, the amplitude, A_R, of the reflected wave will vary in a complicated way with the angle of incidence, but for normal incidence ($i = 0°$) the reflection coefficient (i.e. the ratio of the amplitude of the reflected wave to that of the incident wave) is given by :

$$R_C = A_R/A_i = (\rho_2 V_2 - \rho_1 V_1)/(\rho_2 V_2 + \rho_1 V_1) \qquad (2.9)$$

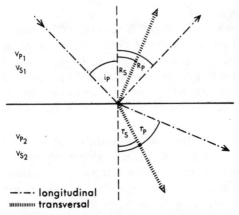

Fig.7.Reflection and refraction of an incident longitudinal wave at a boundary separating two media with different velocities.

Thus, R_C depends on the contrast in "acoustic impedance" (product of density and P-wave velocity) on opposite sides of the boundary. If $\rho_1 V_1$ is greater than $\rho_2 V_2$, R_C is negative and the reflected wave would show a phase change of $180°$; a compression would be reflected as a rarefaction and vice versa. Since in practice both the density and velocity usually increase with depth $(\rho_2 V_2 > \rho_1 V_1)$, most strong reflections are positive, i.e. without phase change. However, in special situations (e.g. salt structures in sediments) the velocity can increase and the density can decrease in such proportion that $\rho_2 V_2 \approx \rho_1 V_1$. In such cases no reflections would be observed. The acoustic impedance is, therefore, a parameter of great importance in seismic reflection studies. The variation in the absolute value of the reflection coefficient R_C as a function of the acoustic impedance ratio $(\rho_2 V_2 / \rho_1 V_1)$ across a boundary is shown in Fig.8:

When a boundary representing a reflecting surface displays a sharp break (e.g. due to a fault or an escarpment), the incident wavefront is diffracted at the sharp corner of the discontinuity. The phenomenon is analogous to that in optics,

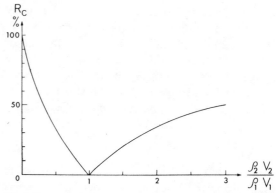

Fig.8.The reflection coefficient (i.e. the ratio of the amplitude of the reflected wave to the amplitude of the incident wave) in the case of normal incidence depends on the contrast in acoustic impedances $(\rho_2 V_2 / \rho_1 V_1)$ separating a boundary. (After Bentz, 1961.)

the sharp corner serving as a centre of scattered waves. If such discontinuities are present in a reflector, they may blur the reflected waves considerably. Diffracted waves have proved to be very useful as indicators of minor faults which are otherwise not easy to detect by the reflection techniques.

When an incident wave strikes a boundary, part of the energy is refracted across the boundary; if the underlying layer has a higher seismic velocity, the wave is refracted towards the boundary (Fig.7). The angle of refraction is given by Snell's law:

$$\sin i / \sin r = V_1 / V_2 \qquad (2.10)$$

From eq. 2.10 it follows that when $\sin i = V_1/V_2$, $r = 90°$. This critical refraction along and parallel to the boundary is of basic importance in the seismic refraction method. It occurs whenever the incident wave strikes the boundary at the critical angle, $i_c = \sin^{-1} V_1/V_2$. The critically refracted wave then travels along the boundary with velocity V_2, but during this propagation the interface is subjected to an oscillatory stress and each point on it sends out secondary waves so that the energy reemerges in the upper layer along rays at angle i_c (again following Snell's law). Such doubly refracted waves give information about the depth of the boundary in cases where a velocity change is involved. It was essentially in this manner that Mohorovičić, from earthquake wave studies, first demonstrated the existence of a "crust" overlying a "mantle" of higher velocity.

Seismic velocities in rocks

In the application of seismic methods to geological problems the most important property of rocks is the velocity of propagation particularly that of longitudinal waves which are the fastest and first to be recorded. We noted earlier that reflection and refraction characteristics of a seismic wave depend primarily on the velocity contrasts involved across the boundary. Thus, a knowledge of the velocities of rocks is basic to seismic interpretation. Seismic velocities of rocks depend on the elastic moduli and density (see eq.2.6) and it is possible to calculate the velocities from measurements of the elastic constants on rock samples. However, the seismic velocities computed from measured elastic constants on samples are considerably different from the velocities of rocks measured in situ. In-situ measurements are, therefore, preferable wherever it is feasible to make them. Various field methods for the determination of seismic velocities shall be described later (p. 55).

The range of variation in the seismic velocities of rocks is considerably greater than the corresponding variations in their densities. Typical values of the longitudinal velocities, V_p, of some rocks listed in Table I should serve to give a rough idea of the velocity range expected for a rock type.

Seismic velocities often show anisotropy in stratified formations. For example, in shales and slates the velocity along the bedding direction may be approximately 10–20% higher than in the perpendicular direction. Also, the effects of compaction

SEISMOLOGY AND THE EARTH'S STRUCTURE

The earth's interior is inaccessible, and we can infer its structure and composition only from indirect evidence. An earthquake involves the sudden release of energy which is transmitted through the earth in all directions in the form of seismic waves. The passage of earthquake waves through the earth provide a kind of X-ray examination of its interior, and these waves give information about the physical structure and properties of the earth's interior. This information, although not conclusive, is by far the most detailed that can be obtained by a geophysical method.

With an earthquake, there is no prior knowledge of the precise origin time and source location. Because these unknown factors have to be derived from the records of seismic waves reaching the surface, there is a consequent lowering of the precision with which details about the earth's interior can be inferred. In contrast, underground nuclear explosions provide point sources of seismic energy with all uncertainties about the origin time and location of source removed. Records of precisely controlled nuclear explosions have been invaluable in extracting finer details besides supplementing the data from earthquakes. In addition, extensive studies made by chemical explosions over various areas, combined with geological information about surface rocks, has greatly increased our knowledge of the outer shell of the earth. A concise review of modern seismology and new developments in the related fields will be found in Båth (1973).

In this section we shall summarize the present knowledge about the earth's structure as inferred from seismological observations.

Earthquake waves and seismogram

The seismic waves generated by an earthquake are usually taken to originate from a single point source within the earth, although in reality the source is often extended over a large

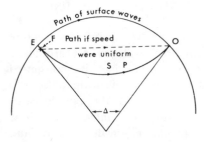

Fig.10.Schematic illustration of the least-time wave paths for P, S, and surface waves from earthquake focus F to seismic observatory O. E is the epicentre and Δ is the epicentral angle.

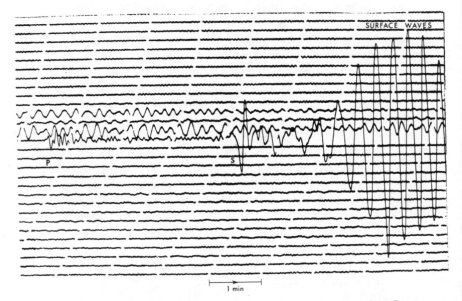

Fig.11. Horizontal seismogram (north component), recorded at Copenhagen, from the earthquake off Iceland, July 23, 1929. (Courtesy of I. Lehmann.)

area. The place inside the earth where the earthquake occurs is called the "focus"* and the place on the surface vertically

*Also called the "hypocentre".

above is the "epicentre". If the velocity of seismic wave propagation (V_p, V_s) were the same for the entire earth, the wave path would be a straight line between focus F and recording station O on the surface. Actually, V_p and V_s are generally observed to increase with depth, and the waves follow curved paths like those shown in Fig.10. Since $V_p > V_s$, P waves arrive at a recording station before S waves, and the time interval between the two arrivals is a measure of the distance to the source. Special tables (see Jeffreys and Bullen, 1967) are used to convert the $S - P$ time interval to approximate epicentral distance.

Fig.11 shows an actual seismogram, the registration of an earthquake recorded at a station near Copenhagen. The registration is characterized by several "bursts" of energy of varying periods corresponding to the wave paths followed by the different waves, many routes being possible because of the reflection at the earth's surface and the reflection and refraction in the interior. The various wave groups in a seismogram are referred to as seismic "phases". The first to arrive is the phase of the direct P wave. This is followed by various other phases of P and S waves. Last to arrive are the surface waves (long waves) which have travelled at relatively slow speeds along the earth's surface. These are generally characterized by large amplitude waves of long periods. It is the task of the seismologist to correctly identify the various phases, and to establish the origin time and location of the earthquake from their arrival times at the different stations. The velocity distribution at depth can then be computed from the travel-time data, although the calculations are rather complicated. The mathematical procedures can be found in Bullen (1963).

Internal layering of the earth

The velocity of the P and S waves have been determined for various depths from the analysis of travel-time data. Fig.12 shows the velocity distribution for P and S waves within the earth according to Jeffreys (1939) and Gutenberg (1959).

24

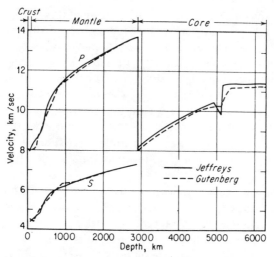

Fig.12.The broad velocity-depth distribution in the earth as deduced independently by Jeffreys (1939) and Gutenberg (1959).

The first major discontinuity (the "Moho") separates the crust from the underlying mantle, at the top of which V_P is commonly about 8·1 km/s, although this velocity ranges regionally from about 7·7 to 8·4 km/s. The depth to the Moho also ranges regionally from an average of about 35 km under the continents to about 6 km below the ocean floor. Another major discontinuity is observed at a depth of about 2900 km. The large drop in V_P from about 13·7 km/s to about 8 km/s (while V_s decreases from about 7·2 km/s to nearly zero) represents the core–mantle boundary, also referred to as the Gutenberg–Wiechert discontinuity. Since no S waves appear to pass through the outer core, it is concluded that the outer core will not support transverse (shear) stresses, and thus it behaves like a liquid. Within the core a rapid increase in P-wave velocity at a depth of about 5100 km was discovered by Miss Lehmann from observations of weak P arrivals at large epicentral distances. These observations led to the subdivision of the core into an outer core (presumably liquid) and a smaller inner core (probably solid).

On the basis of the velocity–depth distribution, Bullen (1963) has divided the earth into a series of shells (see Fig.13) with the boundaries at levels where the velocities, or the velocity gradients, change abruptly. Bullen's subdivisions, although still tentative form the basis for most discussions of the physical and chemical properties of the earth's interior.

The low-velocity layer

The major discrepancies between Jeffreys' (1939) and Gutenberg's (1959) velocity curves (see Fig.12) are within the upper mantle and at the transition to the inner core. In the upper mantle, Gutenberg's velocity distributions show a well-defined low-velocity layer at 100–150 km depth; as early as 1926 he proposed the existence of this layer. The improved techniques of modern seismology have greatly refined the details of the velocity structure in the upper mantle. The existence of a low-velocity zone (LVZ) in the upper mantle appears to be supported by many recent studies (particularly from precisely controlled underground nuclear explosions), although there is less certainty about its depth in relation to tectonic environments. Lehmann (1967) has summarized the evidence from body-wave data. More positive evidence, however, has come from dispersion studies of surface waves.

Fig.14 shows three types of velocity distribution obtained by Dorman (1969) for three different tectonic areas. The pronounced velocity minima associated with the Alpine and oceanic models extend from about 70 to 220 km in depth; the stable shield model displays a less pronounced minimum, although extending much deeper. Important are the variability in depth and the intensity of the minima defining an LVZ in the upper mantle beneath different tectonic areas. Evidently the geological differences between ocean basins, continental shields, and mobile belts are not limited to the crust but extend to a depth of several hundred kilometres into the mantle. It is convenient to adopt the term "lithosphere" (relatively strong

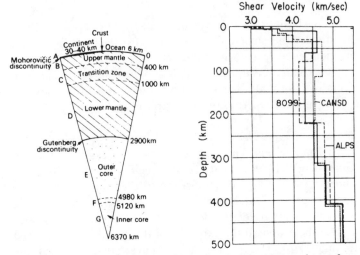

Fig.13.The layering within the earth. (After Bullen, 1963; redrawn from Bott, 1971, fig. 1.6.)

Fig.14.Three shear velocity models for the upper mantle derived from Rayleigh-wave dispersion. All the derived models, CANSD (from the Canadian Shield data), 8099 (from the Pacific Ocean data), ALPS (from the Stuttgart–Oropa data) show a layer with low S-wave velocity. (After Dorman, 1969.)

and rigid zone) for the upper part of the earth, including crust and uppermost mantle, which lies above the low-velocity zone.

The great Chilean earthquake of May 22, 1960, provided further evidence for the worldwide extent of the low-velocity layer. It was so violent that it set the entire earth into vibration, making it "ring" like a bell. It was already known that the earth, being an elastic body, could go into normal mode oscillations ("free oscillations"), but the "ringing" excited by the Chilean earthquake lasted for about a month and was recorded by sensitive seismographs at various observatories. From records of the free oscillations following a big earthquake it is possible, with enormous mathematical labour, to deduce the elastic layering of the earth. The results of this labour show that an LVZ is needed to account for the observed long-period oscillations of the earth. More precisely, the low-velocity plastic layer extends from about 60 km to about 300 km in depth, the actual depth range depending on the tectonic environments.

One possible cause for the existence of an LVZ is the partial melting of rocks under temperature and pressure effects. This could be due to the increased temperature gradient at depths of 50–200 km (Lubimova, 1967), which is caused by the decrease of thermal conductivity. If this is true, the relatively weak and soft region ("asthenosphere"), characterized by a low velocity, takes a very important place in the modern geophysics, especially in the light of the new concept of sea-floor spreading. The partially molten material of the low-velocity layer could provide an ideal medium for the sea floor to glide over. The reality of an LVZ may have an important bearing on the theories of continental drift and convection in the mantle.

Density and other parameters in the earth

The mean density of the earth, estimated from gravitational considerations (p. 90), is about $5 \cdot 5$ g/cm^3 (or $5 \cdot 5 \times 10^3$ kg/m^3 in SI-units)*, which is much greater than the densities of crustal rocks, which usually range between $2 \cdot 5$ and $3 \cdot 3$ g/cm^3. This implies that the earth's interior must consist of much denser material. The essential problem is to discover how the density is distributed inside the earth.

We recall that V_P and V_S are related to the density, ρ, and elastic constants, k and n, of a medium; however, knowledge of only two parameters, V_P and V_S, is not sufficient to extract the three unknowns (ρ, k, n) from two equations (eqs.2.6 and 2.7). In overcoming this difficulty, an indirect procedure (the Adams–Williamson method) has been employed by Bullen

---

*The SI-unit for density is kg/m^3; however, the subunit g/cm^3 which is the same as Mg/m^3 ($= 10^3$ kg/m^3) will be retained in this book, because of its widespread use in geophysical literature.

(1963) to derive plausible earth–density models. Any acceptable model of the density distribution must reproduce the observed values of the earth's mass and moment of inertia. Fig.15 shows one of Bullen's models. Two other density models are also shown for comparison. Birch's (1961) model is based on the linear relationship between V_p (Gutenberg's velocity distribution) and ρ. Anderson's model (1967) is based on the velocity–density relationship, determined from the free oscilla-

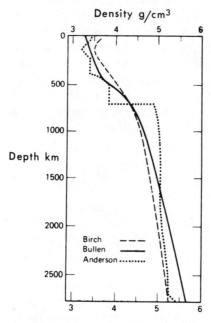

Fig.15.Birch's (1961) and Bullen's (1963) density models for the mantle as compared with Anderson's (1967) model. Note the low-density zones in the models of Birch and Anderson.

tion data. Note the low-density zones in the uppermost mantle as indicated in Birch's and Anderson's models. At least, these models give an indication of the present uncertainty in the density distribution within the mantle. The densities in the upper mantle (to a depth of about 250 km) appear to be in the range $3\cdot35$–$3\cdot6$ g/cm³, which corresponds to the range covered by peridotite to eclogite. Clark and Ringwood (1964) have proposed a "pyrolite"* (hypothetical peridotite) density model for the upper mantle, based mainly on petrological arguments. All models indicate that the density of the core is consistent with a composition dominantly of iron.

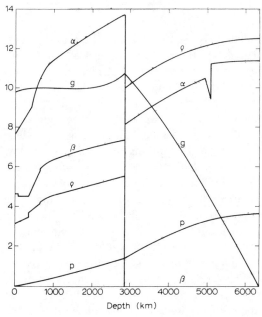

Fig.16. A recent model of the physical properties inside the earth. p = pressure in 10^{12} dyne/cm²; ρ = density in g/cm³; g = gravity in 10^3 cm/s²; α and β are the P- and S-wave velocities, respectively, in km/s. (After Haddon and Bullen, 1969.)

*The term "pyrolite" indicates a mixture of dunite and basalt in the ratio of $3/1$. Fractional melting of this system could provide basaltic magma, leaving a residue of dunite.

The distributions of seismic velocities, density, elastic constants, gravity, g, and pressure, p, within the earth are closely interlocked. Thus, once the distributions of V_p, V_s, and ρ are known, distributions of other parameters (k, n, g, p) can be easily calculated. The present knowledge of these physical properties inside the earth is summarized in Fig.16.

SEISMICITY AND SEISMOTECTONICS

Global seismicity

The term "seismicity" is used to describe the geography of earthquakes, particularly their distribution, frequency, and energy relationships to surface features. Fig.17 shows a world-wide distribution of earthquake epicentres for the period 1961–1967. From the map it is apparent that the earthquake zones are not randomly distributed, but instead, are aligned in belt-like patterns. The circum–Pacific belt of island arcs, deep trenches, and mountain ranges contains about 80% of shallow earthquakes (focal depth $h<70$ km), 90% of intermediate earthquakes (h between 70 and 300 km), and nearly all deep-focus earthquakes ($h>300$ km). Most of the remaining large earthquakes occur in the Alpide belt, extending from the Azores through Europe and Asia to join the circum–Pacific belt in New Guinea. In addition to these two major zones, a significant belt of small shallow-focus earthquakes follows the crest of the ocean-ridge system and extends along the East African rift system. Numerous small events do occur elsewhere, but their energy release is insignificant. Shallow shocks are frequently observed in the vicinity of volcanoes and some of these (but not all) are associated with volcanic explosions.

According to the modern concept of "plate tectonics" (discussed more elaborately in Chapter 8, p. 366), the earthquake belts roughly outline the boundaries of rigid crustal (lithospheric) plates which can move easily on the relatively warmer and softer material of the upper mantle (asthenosphere). The driving mechanism of the plate motions is still debated, although most models envisage some form of convec-

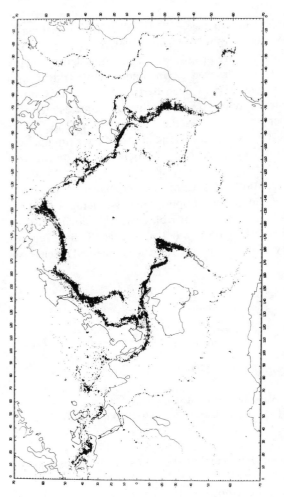

Fig.17.Worldwide distribution of all earthquake epicentres for the period 1961–1967 as reported by the U.S. Coast and Geodetic Survey. (After Barazangi and Dorman, 1969.)

tion in the upper mantle (see p. 378). When the crustal plates collide against each other or slide relative to each other, shallow earthquakes are produced. On the other hand, when a lithospheric plate is dragged down (underthrusted) deep into the mantle, it gives rise to deep earthquakes. For example, in the Tonga arc region (Fig.18) the earthquakes tend to become deeper and deeper as their epicentres extend farther and farther to the west of the island arc until a maximum focal depth of about 600 km is reached. The foci distribution suggests an earthquake zone, dipping at an angle of about 45° down towards the continental side of the arc. Deep-focus earthquake activity along this dipping zone (sometimes referred to as the "Benioff zone") is attributed to the downthrusting of a lithosphere slab deep into the mantle. Strong support for this theory has come mainly from seismic focal mechanism studies.

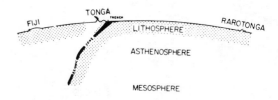

Fig.18.Dipping plane of earthquake foci under the Tonga Island arc. The maximum focal depth reaches to about 600 km. (After Oliver and Isacks, 1967.)

The significance of earthquakes must also be considered in terms of the energy released. Richter (1935) set up a scale of magnitudes for earthquakes that has been of great practical use in establishing a quantitative measure of the elastic strain energy released in an earthquake. The Richter magnitude, M, was defined as $M = \log_{10} A$, where A is the maximum trace amplitude in microns, recorded by a standard Wood–Anderson type short-period seismograph at an epicentral distance of 100 km. Several empirical amplitude–distance tables were devel--

oped to enable observations at distant stations to be reduced to standard magnitude values. The main use of magnitude M is to enable the estimation of the energy, E, associated with earthquakes. An empirical relationship (Båth, 1966) currently in use is:

$$\log_{10} E \text{ (in ergs)}^* = 12 \cdot 24 + 1 \cdot 44 M \qquad (2.12)$$

Using this relationship, the energy associated with one of the largest earthquakes of the century, the Colombia–Ecuador shock of January 1906 with a Richter magnitude of $8 \cdot 9$ is of the order of 10^{18} joules, which is equivalent to the energy released by a 300-megaton bomb. The average total release of energy from earthquakes is of the order of 10^{18} joules/year, and most of it comes from a few really large shocks. Most of the large shocks are known to occur at shallow depths usually within 30 km. The frequency of intermediate and deep shocks generally decreases with increasing depth. No shocks are known to have occurred at depths below 720 km, and it is possible that the deeper parts of the earth are completely aseismic.

Scandinavia's seismic activity

In comparison with other seismoactive zones of the world the seismic activity in Scandinavia is extremely small. This becomes all the more evident from the fact that a single earthquake of large magnitude (of a type which commonly occur in Japan) releases energy of the order of 10^{17} joules, which is a thousand times more than the total earthquake energy released in Scandinavia during this century. Nevertheless, the seismicity of Scandinavia has been of considerable interest, especially in connection with its possible relation to the recent uplift of land in that area, the maximum uplift being recorded around the Gulf of Bothnia (see Fig.63).

*The SI-unit for energy is joule: $1 \text{ J} = 10^7$ ergs.

34

Fig.19 shows the frequency of the earthquake distribution in Scandinavia for the period 1600–1925, the information was compiled from historical records. The most significant activity is localized in three areas, namely along the west coast of Norway, around the Oslo graben system, and in the area adjoining the Bothnic Bay. The seismic activity in Denmark is relatively much less compared to that in Norway, Sweden, and Finland. Here the small earthquake zones are

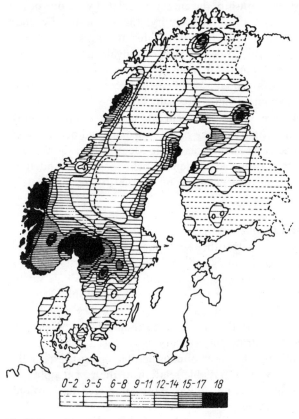

0-2 3-5 6-8 9-11 12-14 15-17 18

Fig.19.Frequency of earthquake distribution in Scandinavia for the period 1600–1925. (After Båth, 1954.)

confined to the salt-dome province in Northwest Jutland and some parts of North Zealand. Obviously, the seismic activity of all these scattered zones cannot be attributed solely to the continued rising of the land since the melting of the Pleistocene ice cap. It is possible that some other factors such as gradual subsidence of the North Sea basin, partial rejuvenation of the "dead" fault system in the Oslo graben, and salt tectonics in the Danish Embayment play a more important role in the build up of local elastic stresses that eventually lead to earthquakes. Before going into the details of the possible causes, let us first try to visualize what exactly happens at the source point (focus) where the earthquake originates!

Earthquake mechanisms

Much of our knowledge of the events that occur in the vicinity of the focus of an earthquake has come from the "elastic rebound" theory, formulated by H. F. Reid after the great San Francisco earthquake of 1906. The theory attributes earthquakes to the progressive accumulation of elastic-strain energy in tectonic regions and the sudden release of the stored energy by faulting when the fracture strength is exceeded (Fig.20). A large shallow earthquake is normally accompanied by substantial deformation (rupture) of the ground over hundreds of kilometres, and this is indicative of the volume of the

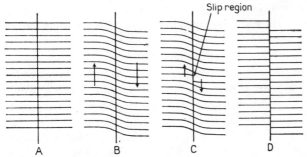

Fig.20.Sketch illustrating H.F. Reid's elastic rebound earthquake-source mechanism. (After Benioff, 1964.)

rock from which elastic strain is released. The disastrous San Francisco earthquake of 1906 was accompanied by a surface rupture of more than 300 km along the San Andreas fault. The displacement of the fault was purely horizontal, the Pacific side moving abruptly north by about 7 m relative to the continental side.

Referring to Fig.20C, if the displacement (slip) is in a horizontal direction, the slip is designated as a "strike-slip", if it is in a vertical or steep direction, it is called "dip-slip". Combinations of dip-slip and strike-slip faulting also occur. The San Andreas fault, the most closely studied fault on the earth's surface, is a typical strike-slip fault. The rate of accumulation of elastic strain energy is known for the San Andreas fault. Recent geodetic measurements show that the fault blocks are moving at the rate of about 2–3 cm/year. This movement, though almost imperceptible, is really large in terms of the geological time scale. Considering the age of the fault (ca. 20 m.y.), this could represent a total movement of the order of several hundred kilometres.

The concept of fracture and sudden faulting implied in the elastic rebound theory is plausible only for shallow shocks. The shear–fracture theory of faulting seems to be inadequate to explain the focal mechanism of deeper events because at depths below some tens of kilometres the frictional resistance is excessively large for any dry frictional sliding to occur. Several solutions to this problem have been suggested, some of which are discussed below.

One possibility envisages a mechanism of a "plastic" nature, involving the sudden collapse of a small volume of rock at the focus by a phase change, particularly fusion. The void would be filled by quick downward movement of the overlying rock, causing an earthquake. For the Peru earthquake of 1966 (focal depth~600 km), strain seismographs indicated a downward motion of the earth's surface at the epicentre. It has been suggested by some that the gradual phase changes in the downsinking slabs of the lithosphere (along the Benioff zone) cause stresses which are released as earthquakes.

In order to explain the deeper shocks another theory claims that the focal mechanism is, in fact, faulting, made possible at depths by the presence of a high-pressure pore fluid that reduces the frictional resistance to slip while leaving the shear stresses unaffected. In relation to this theory, of particular interest are the observations of an apparent increase in local shallow-earthquake activity after high-pressure fluids have been pumped into deep disposal wells. Most rocks in the shallow parts of the crust contain entrapped water and, at depth, water could be released by dehydration processes. Also in the mantle, partial melting of mantle material may yield a pore fluid that behaves in much the same way. The observed concentration of deep earthquakes along what according to the theory of "plate tectonics" are believed to be downsinking slabs of lithosphere material at certain continental margins, suggests also that dehydration reactions may continue to occur even to depths of 700 km.

The problems of physical mechanisms causing intermediate and deep earthquakes are far from solved. But despite these uncertainties, much has been learned about the patterns of displacement at an earthquake focus. This information is of great importance in seismotectonic studies.

Faulting dynamics and seismotectonics

The major feature associated with most earthquakes is a fault plane across which the adjacent material slips. From a study of seismograms it is possible to locate the focus of the earthquake, that is, the point where the initial displacement began. In recent decades ingenious methods have been developed for determining the orientation of the fault that causes the earthquake as well as the slip. For an excellent discussion of these methods and field examples the reader may refer to a review article by Khattri (1973). Only a very brief and elementary discussion can be presented here.

The direction of the first motion of P waves spreading from

38

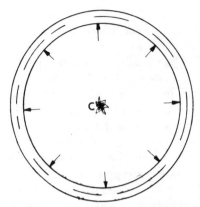

Fig.21.In a homogeneous, isotropic elastic medium a point source of disturbance (*C*) produces a uniform outward pressure (push). The first P-wave arrival is a compression in all quadrants.

a shock centre gives information about the shock mechanism. For example, in Fig.21 the seismic disturbance starting from an ideal point source (such as that provided by a chemical or nuclear explosion) during propagation in a homogeneous elastic medium will produce a uniform outward pressure (push) on the surrounding material in all directions. The first P-wave arrival would be a compression, a push of the particles away from the source, *C*, in all quadrants. The distribution pattern of the first P-wave pulses in the quadrants would be different if the shock mechanism at *C* were due to a linear source such as faulting.

The usual method for studying the displacement pattern at the shock centre (focus) is to observe the sense of the first motion indicated by P-wave arrivals at seismograph stations spread over the earth's surface. Obviously, small earthquakes would be recorded only at nearby stations. For each station, it is noted whether the first P pulse is a compression (push) or a dilatation (pull). The results of all stations are then plotted on a map. Fig.22 shows a hypothetical case of the distribution of first motions associated with a strike-slip movement along a N–S striking fault in the vertical plane *FF'*. The first

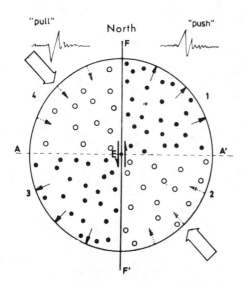

Fig.22. First-motion record of P-wave arrivals due to an earthquake movement along a N–S striking fault. E=epicentre, FF'=trace of the fault plane,
AA'=trace of the auxiliary plane, • =compression, o=dilatation; the
arrows indicate the directions of maximum horizontal stress.

motions as registered by seismograph stations can be divided
into four quadrants. The quadrant-type distribution of compressions and dilatations would be exactly the same, if the
strike-slip fault were to be E–W oriented along the auxiliary
plane AA'. Although the direction of the maximum horizontal
stress (MHS) can be uniquely established from the distribution
of first motions towards the epicentre, there are two alternative
solutions for the direction of the fault plane, and we do not
know solely from the P-wave first-motion studies which of the
two planes (FF' or AA') represents the actual fault plane.
Regional geology may be useful in resolving this ambiguity,
or a solution based on S-wave data may be of significance
(Chandra, 1970). Sometimes macro-seismic indications, such as
isoseismal lines (curves of equal seismic intensity made from

40

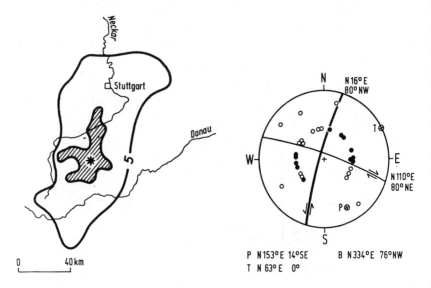

P N153°E 14°SE   B N334°E 76°NW
T N 63°E 0°

0   40 km

Fig.23.Fault-plane solution of the earthquake in the Swabian Jura area, South Germany, January 22, 1970 ($M=5.3$). Two possible solutions are obtained from the analysis of the first P-wave arrivals. The isoseismal curve (thick line) around the epicentre indicates that the faulting plane has a NNE–SSW direction. *=epicentre, ●=compression, o=dilatation. (After Ahorner et al., 1972.)

observed disturbances around the earthquake epicentre), provide a clue for determining the strike of faulting.

As an example, Fig.23 shows the actual observations from the January 1970 earthquake in the Swabian Jura area of South Germany. The projection diagram shows that the first pulse is compressional in the ENE and WSW quadrants and dilatational in other quadrants. The arrows are added to be consistent with the compressions and dilatations. The isoseismal curve around the epicentre clearly indicates that the faulting plane is in a NNE–SSW direction. From the fault-plane solution we can infer that the movement is almost horizontal and left-lateral (sinistral). The direction of the maximum horizontal stress (MHS) as determined from the first-motion studies is found to be roughly NW–SE which agrees with the mean direction of the tectonic pressure axis in the Rhine Graben area.

Fig.24 shows the seismotectonic stress field in the north-western part of Central Europe, as deduced from first-motion studies of earthquakes. The great majority of strike-slip shocks in the Rhine area and its surroundings indicate a mean direction of maximum horizontal stress (tectonic pressure axis) in NW–SE, and a mean direction of the least horizontal stress (tectonic tension axis) in SW–NE. Under the influence of this widespread regional stress field the diagonally arranged fracture zone of the Upper Rhine Graben is affected by horizontal shear forces which try to move the eastern block relatively to the north and the western relatively to the south. Something

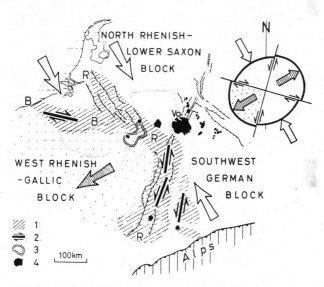

Fig.24.Seismo-tectonic scheme of the northwestern part of Central Europe. Arrows indicate the present stress field and possible block movements. 1 = earthquake zones (R = Rhenish zone, B = Brabant zone), 2 = seismic active wrench-faults, 3 = area with Quaternary volcanism, 4 = main centres of Tertiary volcanism. (After Ahorner, 1970.)

analogous happens to its structural counterpart, the WNW–ESE striking pattern of the Brabant seismo-active zone. Along this zone, which marks approximately the boundary between the Caledonian Brabant massif and the Hercynian Rhenish massif, right-hand horizontal shear movements take place and shift the northern block relatively to the east and the southern relatively to the west. The proposed model of horizontal block movements in Fig.24 would explain the observed tectonic features, e.g. the existence and the trend of the Rhenish and Brabant seismic active zones and the occurrence of active wrench-faults within these zones. For a detailed discussion of regional seismotectonic studies in Europe and elsewhere the reader is referred to the very interesting papers by Ahorner et al. (1972), Pavoni (1969), and Wollard (1969).

The main value of first-motion studies is to determine the direction and type of movement associated with earthquakes and the distribution of the principal stress axes in tectonic zones. The technique has proved to be a very powerful tool in seismology for gaining valuable information on the present tectonic activity of the earth's crust. Probably the most extensive recent application of the method is that of Isacks et al. (1968). The earthquakes used were those associated with ocean margins and mid-ocean ridges, and the results are of great significance in the study of ocean-floor tectonics. Such studies conclusively show that the transverse faults on the Mid-Atlantic Ridge are "transform faults" (Fig. 184, p. 365) with a sense of movements consistent with the hypotheses of sea-floor spreading and plate tectonics. We shall elaborate this point later in Chapter 8.

METHODS OF SEISMIC PROSPECTING

Seismic prospecting has as its main objective the mapping of the geological structures almost invariably in the uppermost part of the earth's crust. The general theoretical background of seismic prospecting has much in common with earthquake seismology, the principal difference being in the period or frequency of the seismic waves used. In contrast to the earthquake waves, the waves used in seismic prospecting are of very short wavelength with periods ranging from 0·01 to 0·1 second, the corresponding frequencies being in the range of 10–100 cycles per second. The fundamental principles of the instruments used are generally the same, but their technical specifications and operational details are greatly different.

In seismic prospecting, the basic procedure is to generate seismic waves by a near-surface explosion, to record the resulting waves which reach the surface at various distances through different paths, and to deduce the positions of reflecting and refracting interfaces by analysis of the travel times of identifiable wave groups. The techniques using reflected waves differ considerably from those based on refracted waves and the former have been most extensively used in oil exploration for mapping of structural traps.

An introductory account of seismic prospecting methods has been given by Dobrin (1960). Details of instrumental requirements and field routine can be found in a recent monograph by Anstey (1970). Techniques of seismic interpretation have been discussed by Dix(1952), Gurwitsch(1970), and Gassmann(1972).

The following is a general discussion of the subject, the relative emphasis being on the principles involved rather than the technical details.

Generation, detection and recording of seismic waves

Whatever the method (refraction or reflection), a device for generating and detecting seismic pulses is essentially needed. By far the most common method of producing seismic waves is to explode a dynamite charge in a hole, the amount of charge depending on the depth of the geologic structure to be investigated. In contrast to the refraction studies, where source–detector distances are comparatively very large, the reflection work requires a lesser amount of explosives. The main drawback of explosives is that, despite all the safety precautions that are required in their handling, they cannot be used in densely inhabited areas. Other methods of generating seismic waves have been tried, e.g. weight dropping (patented after **B.** McCollum, 1956), powerful electric or gas sparking*, etc. For shallow-depth investigations (e.g. in civil engineering and foundation problems) sledge-hammer blows on the ground often serve as an adequate source of energy. In off-shore seismic work, an electric or gas spark fired under water is commonly used as an energy source. Recently some vibratory sources (e.g. VIBROSEIS, a trademark of the Continental Oil Company) which continuously generate short-duration pulses have been tried successfully in geologically diverse areas. The vibratory method is inherently much safer and more convenient than a method using explosives.

The seismic disturbance, beginning from the source and travelling along different paths, is detected on reaching the ground surface by an array of "geophones". A geophone is a type of microphone devised to "listen" to the minute ground vibrations (as small as 10^{-10} m) of the earth. It consists of a coil and a magnet, one rigidly attached to a frame and the other suspended from a fixed support by a spring (Fig.25). Any displacement of the ground causes a relative motion between the coil and magnet so that an oscillatory voltage is generated in the coil which is proportional to the velocity of the motion. This signal can be amplified and undesirable frequencies can be filtered out. The natural frequencies of geophones used in reflection seismic work are usually 30 c/s or higher, whereas for long refraction profiles they may be as low as 2 c/s. In marine seismic surveys pressure-sensitive seismometers or "hydrophones" are used; these are suspended in the water with the aid of plastic floats. A diaphragm may convert pressure changes to mechanical motions detected in a way similar to the geophone described above, or piezoelectric

---

* The airgun with a mechanical air compressor is at present the most successful tool for marine reflection seismic work.

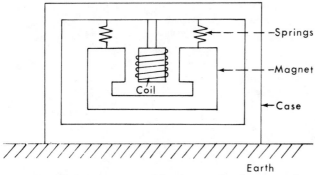

Fig.25. Simplified sectional view of the electromagnetic type geophone. Here the coil, attached to the case, moves with the earth while the magnet acts as the inertial element.

transducers may convert the changes in pressure directly into electrical impulses.

The electrical output from the geophone (or hydrophone), after suitable amplification and filtering, is fed into a recorder unit where it sets a tiny galvanometer into oscillations. The galvanometer oscillations can be registered by means of a mirror-light source system on a continuously running photographic film*. Vertical time lines, generally crossing the entire width of the film, are superposed on the record in order that events such as reflections and refractions can be timed accurately (within a few milli-seconds). Modern multi-channel recorders can simultaneously register signal traces from as many as 48 geophones (3 are shown in Fig.26). With the development of magnetic tape recording of signal impulses, the conventional optical recording has now become almost obsolete except for some small-scale refraction seismic work. The magnetic recording allows greater flexibility because of the play-back feature. Replaying the tape is similar to firing the shot again. This makes it possible to try different techniques of filtering

*An alternative to the photographic film is a special "self-developing" paper (used e.g. by ABEM in the Trio Refraction unit).

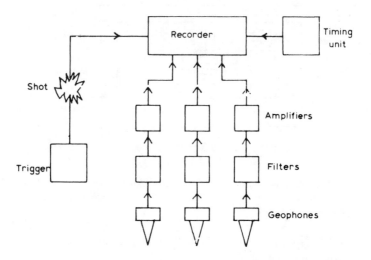

Fig.26. Block diagram of seismic equipment. Modern seismic units have 24 to 48 channels. (After Parasnis, 1973.)

to improve the signal to noise ratio. In fact the magnetic recording system is so sophisticated that all the necessary improvements and corrections in the data can be made during the replay of the tape (Fig.27). For the purpose of interpretation, however, a visual record is always needed, and this is made either in the form of "wiggle" traces (as in Fig.27) or as a "variable-density" or "variable-area" record (see Fig.30). The essential difference between the variable-density and variable-area sections is that in the former the peaks of the conventional record (see Fig.27) appear as black and troughs as white, whereas in the latter the widths of the black and white portions are proportional to the signal amplitudes.

The reflection method

The reflection seismic method has been used far more than any other geophysical method for the mapping of underground structures in the sedimentary section.* The underlying principle

*Particularly in connection with oil exploration.

Fig.27.Magnetic **playback** device. The seismic impulses are recorded on a magnetic tape that can be replayed for further analysis of the seismogram. (Courtesy of Prakla-Seismos, Hannover, W. Germany.)

is very simple and essentially similar to that of echo-sounding, although its practical application can become extremely involved. For reasons, which will be discussed later, the method **is less suitable for deep crustal studies, its optimum practical** applicability being confined to studies at intermediate depths from some hundreds to a few thousands of metres. It is only possible here to give a brief outline of the method and to discuss some of its applications.

Method of reflection profiling. Fig.28 shows one of the sim-

48

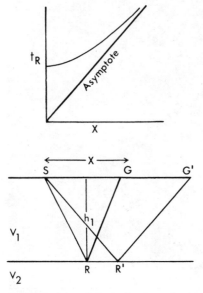

Fig.28.Principle of the reflection seismic method. From the knowledge of the reflection times recorded at geophones G and G', the velocity, V_1, and the depth, h_1, to the interface can be calculated. The upper part of the figure shows the plot of the reflection time, t_R, versus shot-to-geophone distance, x.

plest geological situations where a single horizontal interface separates two rock formations of different velocities. A wave beginning from a shot S is reflected at a point R of the interface and arrives at the geophone G. If h_1 is the thickness of the upper layer and t_R is the time the reflected wave takes to arrive at G, then:

$$t_R = \frac{2\,SR}{V_1} = \frac{2}{V_1}\sqrt{(h_1^2 + x^2/4)} \qquad (2.13)$$

or:

$$h_1 = \frac{1}{2}\sqrt{(V_1^2 t_R^2 - x^2)} \qquad (2.14)$$

The plot of the reflection time, t_R, versus shot-to-geophone distance, x, is a hyperbola convex towards the x-axis and symmetrical about the time axis (Fig.28).

The velocity V_1 of the upper layer and depth h_1 to the interface can be readily obtained by recording the reflection times at two distances (x_1, x_2) and making use of eq.2.14.

The information obtained by a single-reflected pulse at one detector (geophone) position is not sufficient to establish the existence of a reflecting horizon. In practice, an array of geophones $(G, \ldots, G')$ is placed at a relatively short distance* $(x \ll h_1)$ from the shot. The extent of subsurface reflector mapped by the geophone spread GG' would be RR' as shown in Fig.28. By a stepwise shifting of the entire shot–geophone spread configuration, a continuous mapping of the reflecting horizon is possible. The geophone spread length GG' for most reflection surveys is of the order of a few hundred metres with a spacing of some tens of metres between the geophones. As a coarse rule-of-thumb, the spread length GG' is unlikely to be larger than the depth, h_1, to the shallowest reflector. For continuous coverage of the reflecting horizon it is common to use a "split-spread" arrangement with an array of geophones on either side of the shot point.

If there are two or more reflecting horizons which separate layers with different velocities (Fig.29), the corresponding times for nearly vertical reflections are given by:

$$t_1 \simeq 2\, h_1/V_1 \tag{2.15}$$

$$t_2 \simeq 2\, (h_1/V_1 + h_2/V_2) \tag{2.16}$$

and so on for $t_3, \ldots, t_n$.

If the average velocities $V_1, V_2, \ldots, V_n$, for the respective layers are known then the thicknesses, $h_1, h_2, \ldots, h_n$, of the layers

---

*In wide-angle reflection shooting for deep seismic sounding, x is much larger than h_1.

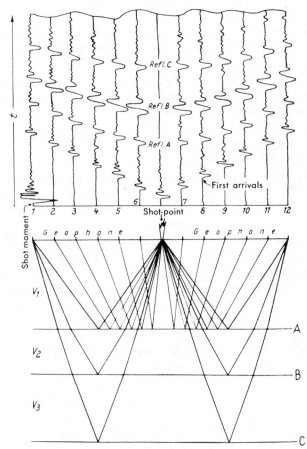

Fig.29.Schematic diagram showing the production of a reflection seismo-gram. The 12-track record shows the time sequence of the reflected pulses from reflecting horizons *A*, *B*, and *C*. For the significance of the first arrivals see Fig.35. (After Bentz, 1961.)

can be calculated. The times, $t_1, t_2, \ldots t_n$, can be read from the reflection record (seismogram). Part of a reflection seis-mogram with 12 traces is shown in Fig.29. A wiggle on a single trace may represent a reflected signal or ground move-ment due to "noise", and there would be no means of distinguishing a signal if only one trace were recorded. However, noise pulses are unlikely to be exactly in phase at all the geophones, whereas reflected pulses from a lithologic interface arrive approximately in phase at the various geophones almost simultaneously since the geophone spread length is relatively small, Distinct reflecting horizons are recognizable by a characteristic "lining-up" of signals of large amplitude across the entire width of the seismogram. The lining-up is almost straight for the later reflections, but for the earlier reflections the pulses lie on a slightly curved line which is actually the hyperbola represented by eq.2.13.

Continuity of the reflecting horizons can be ascertained by the time correlation of a series of seismograms obtained from continuous reflection profiling. Fig.30 shows a visual display (variable-area type) of correlated time sections of the various reflecting horizons in a sedimentary basin of the North Sea area. It is evident that such time sections give a most direct picture of the subsurface geologic structures.

Reflection studies of dipping beds. In a geological section many lithologic interfaces with varying attitudes (dips) may be involved. If reflections are received from a plane dipping interface at a number of points along the surface, the angle of the dip can be determined from the difference in time between reflections. Fig.31 shows the reflection paths followed by waves reaching the ground surface at various detector positions. To simplify the geometry the shot point, S, can be placed at its "mirror image" position, S' (from optical analogy). With this geometry the travel time of a reflection to a surface point at distance x in up-dip and down-dip directions would be:

$$V_1^2 t_u^2 = 4h^2 + x^2 - 4hx \sin \phi \qquad (2.17a)$$

$$V_1^2 t_d^2 = 4h^2 + x^2 + 4hx \sin \phi \qquad (2.17b)$$

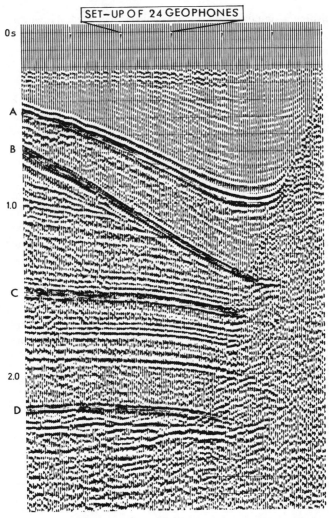

0 s

A

B

1.0

C

2.0

D

SET−UP OF 24 GEOPHONES

Fig.30.Variable-area record of correlated time sections (in seconds) of various reflecting horizons in a sedimentary basin in the North Sea. The horizons marked *A*, *B*, *C*, and *D* represent the base of Tertiary, Upper Cretaceous, Triassic, and Permian formations respectively. Note the abrupt discontinuity in the reflecting horizons (right) due to salt uplift. (Courtesy of Prof. T. Sorgenfrei.)

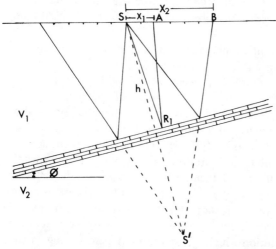

Fig.31.Method of mapping a dipping reflector. The shot point, S, is laid midway between a split-spread array of geophones. From the difference in reflection times recorded at two geophones equidistant from S in up-dip and down-dip directions, the dip angle ϕ can be calculated.

By shooting both "up-dip" and "down-dip" at the two distances, x_2 and x_1, and determining the corresponding travel times, t_u and t_d, from the seismograms, the velocity V_1 and the angle of dip, ϕ, can be determined from the following relations:

$$V_1^2 = \frac{2\,(x_2^2 - x_1^2)}{(t_d^2 + t_u^2)_{\text{at } x_2} - (t_d^2 + t_u^2)_{\text{at } x_1}} \qquad (2.18)$$

and:

$$\sin \phi = \frac{V_1\,(t_d - t_u)_{\text{at } x_1}}{2x_1} = \frac{V_1\,(t_d - t_u)_{\text{at } x_2}}{2x_2} \qquad (2.19)$$

It should be noted that the dip calculated from eq.2.19 gives the component in the direction of the geophone spread. However, if the dips are determined from two shot-geophone spreads at right angles to each other, the true dip and actual strike of the reflecting bed can be determined (e.g. see Nettleton, 1940, p.294).

The calculation of perpendicular depth h to the reflector follows from the eq.2.17 since V_1 and ϕ are already determined. Alternatively, graphical procedures (Slotnick, 1959) can be used to plot the position of a segment of the reflector. The usual approach is first to locate the mirror image, S', of the shot point by drawing arcs of lengths $V_1 t_A$, $V_1 t_B$, etc., from various surface points such as A, B (Fig.31). The point below the surface where the arcs intersect locates the position of S'. The perpendicular bisector of line SS' then gives the position of the reflector segment. The graphical method is versatile because a prior determination of V_1 is not essential. Various solutions for locating S' can be attempted with different assumed values for V_1, and the solution which gives a minimum scattering in the position of S' suggests that the velocity assumed is the correct one.

The case of an n-layer problem with arbitrary dips for reflecting interfaces has been theoretically solved by Gassmann (1961).

It should be pointed out that in practice conditions are much more complicated than the ideal premises upon which the above treatment is based. There is usually a "weathering layer" near the surface, in which the velocity will be different from that at depth, which will cause refraction and which will also affect the calculated depth. In most cases it is advantageous to drill a shot hole in order to put the explosive shot below the weathering layer. The surface topography involving elevation differences between the shot and detector positions will affect the travel times which should be adequately corrected for. The usual procedure is to reduce the reflection times to a "datum-plane" below the weathering layer. Datum-correction methods have been described by Dobrin (1960).

Determination of seismic velocities. To convert the corrected reflection times into depths it is necessary to know the average velocities of the formations lying between the various reflecting interfaces. In a sedimentary formation the velocity is likely to change continuously with depth (generally increasing according to the Faust relationship, see p.19) so that the velocity is not a simple, uniform quantity. Several procedures can be employed for determining the average velocity.

One procedure is to use the relationship between the reflection time, t_R, and the shot-to-detector distance, x. From eq.2.13, it is evident that if we plot t_R^2 versus x^2, we should have a straight line with a slope $1/V_1^2$ and an intercept $4h_1^2/V_1^2$. Thus, the slope of such a curve should give the velocity. Extending this approach to large shot–detector distances, all reflections on the record can be used for this graphical computation and the separate average velocities, V_1, V_2, V_3, etc., can be determined for the respective reflections coming from depths h_1, h_2, h_3, etc. The method requires that the detector spread is shot from both ends and the apparent velocities so obtained are averaged to minimize the effect of dip in reflecting formations.

If deep boreholes or wells are available in the area, the technique of "well-shooting" or that of "continuous velocity logging" (CVL) is invariably used for velocity determinations. The former procedure consists of shooting dynamite charges in a shallow hole alongside a deep borehole in which a detector records the arrival times of the waves at various depth intervals. In the CVL method seismic velocity is measured by lowering an acoustic-pulse producing device in the borehole, its signal is picked up by a closely placed detector. Practical details of this technique (sonic log) can be found in Summers and Broding (1952).

CVL is now part of the established routine in all exploratory boreholes and has many applications, such as geological correlation, identification of formations, and synthesis of reflections, i.e. the making of synthetic seismograms. A part of a CV-log is shown in Fig.32 together with the correspond-

56

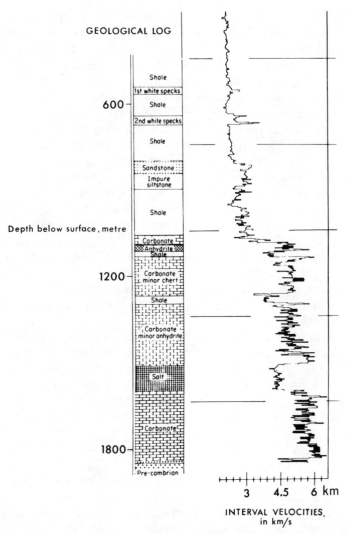

Fig.32. A continuous velocity log with the corresponding geological section. Note the marked increase of velocity shown by carbonate formations. Also note the frequent changes in velocity that occur within the same formation. (After Grant and West, 1965, p. 147.)

ing geological section. The very frequent variations in velocity within some of the formations suggest that the lithology may change significantly within a single formation every few decimetres or so. Since every lithologic change involves a change in elastic properties, the complexity of the reflection processes in a geological section is clear. Lateral changes in velocity may explain why many reflections do not persist for large distances. For a deeper insight into the physical and geological aspects of the reflection process the reader is referred to an excellent article by Woods (1956) and to a general discussion by Dobrin (1960, p.151).

Interpretation of reflection data. Despite the superficial simplicity of the basic principles involved, the analysis and interpretation of the reflection data require great skill. Besides the subjectivity involved in selecting the reflections from a seismogram, there are other factors which complicate the interpretation. As is evident from CV-logs, lithologic changes occur very frequently, and within a single formation many reflectors exist, some large and some small with respect to their reflection coefficients. The relatively frequent occurrence of small reflections adds to the background noise and this is by no means entirely random. In addition, subsurface irregularities (sharp edges, buried erosional features, etc.) in the reflectors may cause waves to be scattered which sometimes mask the reflected events. Furthermore, it is not uncommon to find "multiple reflections" (Fig.33) on seismograms. These pulses have undergone one or more additional reflections either from the surface of the earth or from a lower interface. When recognized and identified, multiple reflections can aid the interpretation; otherwise they complicate the interpretation by introducing spurious reflecting horizons.

Thus, in favourable conditions, when the noise problem is not too severe, if the geologic structure is not too complicated, and if distinctly identifiable reflections can be selected from the field seismograms, then information regarding the subsurface

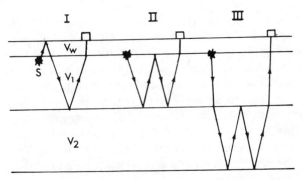

Fig.33.Some possible multiple reflections. Ray paths are shown from a shot S fired below the weathering layer V_w.

position of lithologic interfaces can be derived.

The interpretation procedure consists of selecting reflections from the seismogram and correlating them by juxtaposing of seismograms from consecutive shot–geophone set-ups. For a first rough interpretation only the major features are of interest, and so the variable-area (or variable-density) type cross sections (Fig.34A) serve the purpose adequately.

For a detailed interpretation the reflection time sections are to be converted into depth sections. The process of assigning depths to reflecting horizons on the basis of the known average velocities is called "migration". In current practice, this process is largely carried out automatically by elaborate techniques. Fig.34B shows an example of the migrated section together with its geological interpretation.

As to the reliability of the interpreted data some points are to be noted. In practice the reflected pulses have a frequency spectrum usually in the range of 15–75 c/s. The average velocity in the sedimentary section is likely to be between 2000 and 4000 m/s. If we take typical values, e.g. 30 c/s for the frequency and 3000 m/s for the velocity, this will correspond to a wavelength of 100 m. This would mean, so to say, looking at the reflector with a light, which has a wavelength of about 100 m. With such long reflected-pulse wavelengths, it is

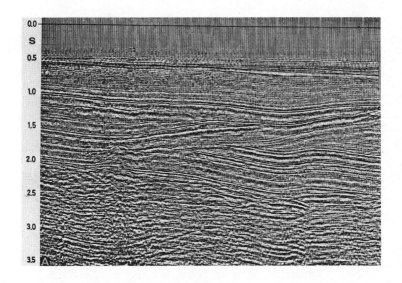

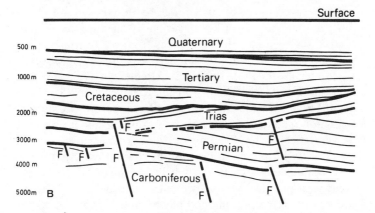

Fig.34.A. Variable-area time section made by juxtaposing of 16 reflection records, each shot with a geophone spread of 2400 m. The reflection times of the various horizons can be read on the vertical scale (in seconds).
B. Migrated (depth) section corresponding to the time section (A) and its geological interpretation by Dr.A.A.Fitch. F denotes the possible location of a fault. (After Anstey, 1970.)

obvious that we cannot expect to resolve variations smaller than 100 m or so. Therefore, small faults, buried erosional features, and stratigraphic features in general are likely to evade the reflection seismograph.

The most successful use of the reflection method has been in the mapping of structural features at intermediate depths, of between some hundreds of metres to a few thousand metres. In the shallower range the resolution problem becomes serious and in the deeper range the increased complication of the structure and the effects of curvature in reflecting beds tend to render the interpretation more difficult and less reliable. For such depth ranges the refraction method may be more advantageous to work with.

The refraction method

The basis of the refraction method is the application of Snell's law of optics to seismic waves. The method has since long been used in earthquake seismology for studying the earth's internal layering as discussed before (p.23). In refraction prospecting, the subsurface layering is detailed on a much smaller scale, using refracted waves from near-surface explosions. The "refraction technique" was first developed by L. Mintrop who in 1919 had it patented in Germany and later founded the first seismic exploration company "SEISMOS". In the earliest days of seismic prospecting, the refraction method was the only one available and it proved to be very effective for locating salt domes. There is now much less activity in refraction prospecting than in reflection. However, the method is particularly valuable for reconnaissance in areas with practically unknown subsurface geology. In recent decades the refraction method has become a powerful tool of the explosion seismology for investigating the crustal structure. On a much smaller scale, the method has also been successfully employed in geotechnical and engineering surveys to determine the depth to the bedrock.

The fundamentals of refraction of seismic waves have been

discussed earlier (p.15). The following is a brief discussion of the application of the method to geological problems.

Mapping of horizontal interfaces. Let us consider a simple case of two layers with velocities V_1 and V_2 ($V_2 > V_1$), separated by a horizontal interface at depth h_1 (Fig.35). When an explosion occurs at S, the energy moves outward in all directions. An incident wave following the path SB and striking the interface at a critical angle i_c ($= \sin^{-1} V_1 / V_2$) is refracted along the interface. The wave then travels along the interface with velocity V_2 of the lower layer, but it sends part of the energy into the overlying layer along waves such that the angle of emergence equals the angle of incidence i_c. There are an infinite number of such emerging waves from the boundary (e.g. BB', CC', DD', etc.), of which the first one, BB', is actually a critically reflected wave (of prime importance in wide-angle reflection shooting).

If we place a number of geophones along a straight line from the shot point, S, the first energy to arrive at the nearer geophones will be that of the direct wave travelling along the ground surface with velocity V_1. However, at the more distant geophones the first wave to arrive will be the refracted wave (also called the "head wave") because it travels part of the path with higher velocity V_2 and thus overtakes the direct wave. Therefore, if we plot the first-arrival times against the shot-to-geophone distances (Fig.35) the first few arrival times will fall on a straight line and the rest on another straight line, the slopes of the lines being $1/V_1$ and $1/V_2$, respectively.

The depth, h_1, to the interface can be determined from the "break-point" distance, x_c, where the two time segments intersect. The relationship between x_c (usually referred to as the crossover distance) and h_1 is easy to derive (e.g. see Dobrin, 1960, p.73) and is given by:

$$x_c = 2h_1 \sqrt{(V_2 + V_1)/(V_2 - V_1)} \qquad (2.20)$$

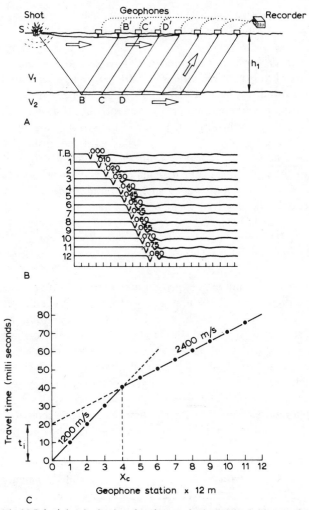

Fig.35.Principle of seismic refraction method. A. The field setup for record-ing the critically refracted waves. B. The refraction seismogram. C. The travel-time plot showing the travel time versus the geophone station distance.

Alternatively, the depth can be determined by using the following relationship between the intercept time, t_i (corresponding to the second segment in Fig.35), and h_1:

$$t_i = 2h_1\sqrt{(V_2{}^2 - V_1{}^2)}\Big/V_1 V_2 \qquad (2.21)$$

It follows from eq.2.20 that x_c is always greater than $2h_1$. Velocity contrasts normally involved in geologic sections are such that $x_c \gtrsim 3h_1$ in most of the cases. This point has to be remembered when planning a refraction survey, for unless the geophone spread is extended to a distance beyond x_c no refracted waves would show up as first arrivals corresponding to the second segment of the time–distance graph shown in Fig.35. The refracted waves occurring within the crossover distance, x_c, would show up only as later arrivals, which can be detected sometimes on the seismogram by another line-up of signals following the first arrivals. In difficult field conditions, where increasing the shot–geophone distance beyond x_c may not always be feasible, second-arrival refraction technique can be used to map the interface. An example of refraction work using second arrivals for mapping the basement below a thick sedimentary section in an area of West Greenland is given by Sharma (1973).

In principle the two-layer interpretation method can be extended to any number of layers with velocities V_1, V_2, V_3, ..., V_n, as long as $V_{n+1} > V_n$. For example, for a three-layer case there will be three segments on the time–distance graph from which the velocities, V_1, V_2, and V_3, can be determined. The depths to the interfaces (h_1, h_2) can be determined from the intercept times corresponding to the respective segments. The necessary formulae can be found in Dobrin (1960) and Gassmann (1972).

Mapping of dipping interfaces. If a refraction profile is shot from both ends of the geophone spread, it is possible to determine the dip of the refracting interface. Fig. 36 shows a case

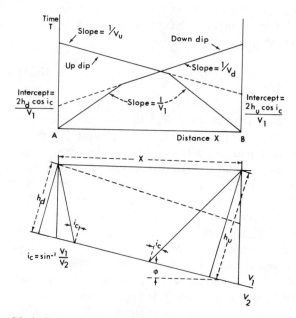

Fig.36.Schematic illustration of critical refraction along an interface dipping at angle ϕ. The two sets of travel-time sections correspond to up-dip and down-dip shooting at the end of profile AB. (After Dobrin, 1960.)

where the interface between two layers is dipping at an angle, ϕ, with the horizontal. Up-dip and down-dip shooting of profile AB would give time segments for each direction of shooting. The inverse slope of the first segment in either case would give V_1, the velocity in the upper layer. In contrast to this, the inverse slope of the second time segment for each direction of shooting would give an apparent velocity (V_d or V_u) which is related to the true velocity, V_2, of the second layer by the following equations:

$$V_d = V_1/\sin (i_c+\phi) = V_2 \sin i_c/\sin (i_c+\phi) \qquad (2.22)$$

$$V_u = V_1/\sin (i_c-\phi) = V_2 \sin i_c/\sin (i_c-\phi) \qquad (2.23)$$

where i_c is the critical angle given by $\sin^{-1} V_1/V_2$.

Solving for ϕ and i_c we obtain:

$$\phi = \tfrac{1}{2} (\sin^{-1} V_1/V_d - \sin^{-1} V_1/V_u) \qquad (2.24)$$

$$i_c = \tfrac{1}{2} (\sin^{-1} V_1/V_d + \sin^{-1} V_1/V_u) \qquad (2.25)$$

The dip is directly determined from eq.2.24. The dip thus calculated is in the direction of the shot–geophone line AB. If another geophone spread is placed perpendicular to the one shown in Fig.36 and shot from both ends, the other component of the dip may be determined in a similar way. In this manner we can determine the total dip, its direction and its strike.

With i_c determined from eq.2.25, V_2 is readily obtained from the relation $\sin i_c = V_1/V_2$. The perpendicular depths h_u and h_d can be determined from the respective intercept times and are:

$$h_u = \frac{V_1 T_{iu}}{2 \cos i_c}; \qquad h_d = \frac{V_1 T_{id}}{2 \cos i_c} \qquad (2.26)$$

Fig.37 shows an example of the travel-time graph obtained by normal and reverse shooting of a refraction profile in the Itivdle valley of Nûgssuaq, West Greenland. The first segments give the velocity (1780–2250 m/s) of the Quaternary sediments and glacio-fluvial deposits flooring the Itivdle valley. The second segments give the apparent velocities (V_u, V_d) of the underlying refractor (Cretaceous sediments), which have an average velocity of approximately 3000 m/s. Using eq.2.26, the depths to the top of the refractor are determined to be about 140 m and 90 m on the eastern and western ends of the profile, respectively. The apparent eastward dip (about 3°) for the refractor suggested by the above depth figures may not be representative of the true dip of the Cretaceous sediments, as the variations in depth in this case could have been caused by the erosion of the valley.

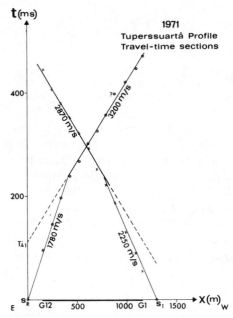

t(ms)

1971
Tuperssuartâ Profile
Travel-time sections

Fig.37.Refraction travel-time plot of a reversed profile in Nûgssuaq. S_R and S_1 are the shot points and G_1 G_{12} is the geophone line.

The case of an n-layer refraction problem with an arbitrary dip for each interface has been dealt with by Gassmann (1972). Nomograms for the rapid determination of interface depths and dips have been published by Meidav (1960) and Habberjam (1966).

Effect of uniform increase of velocity with depth. An important point in relation to refraction surveying in sedimentary basins is the continuous increase of velocity with depth. The effect of depth on velocity is expected to be larger for younger rocks, which may still be undergoing compaction, while older rocks which have suffered a long history of vertical movements and compaction, may possess velocities largely independent of the present depth of burial. Common rates of the increase

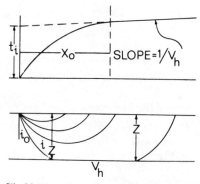

Fig.38. Wave paths and time–distance curve for a sedimentary section, where the velocity increases linearly with depth, overlying a high-velocity layer (V_h). (After Nettleton, 1940.)

of velocity with depth found in sedimentary sequences range from about 0·3 to 1 m/s per metre.

A geologic section in which the velocity increases continuously with depth (Fig.38) is likely to cause problems of depth penetration in refraction studies. These problems will occur because of the curved ray path followed by the refracted wave. For a uniform increase of velocity with depth ($V = V_0 + kZ$), the maximum depth of penetration can be expressed by the following formula (see Nettleton, 1940, p.258):

$$Z_{max} = \sqrt{(X/2)^2 + (V_0/k)^2} - V_0/k. \qquad (2.27)$$

where X is the shot–detector distance, V_0 the velocity at zero depth and k a constant.

Taking a typical set of values for V_0 and k as 2400 m/s and 0·5 m/s per metre respectively, Table II shows the depths of maximum penetration as a function of the shot–geophone distance X. It will be seen that there is no significant penetration for a curved ray trajectory for shot–geophone distances of the order of 5 km. A deep sedimentary basin with a monotonic increase of velocity with depth would not be a promising object for refraction studies. In oil exploration, refraction survey is useful only where there is some definite discontinuity in seismic velocity that can be detected or mapped. This condition may exist in geological situations where a continuous series of sands and shales is deposited on a crystalline basement or over a massive limestone with a relatively high velocity.

Mapping of faults and salt domes. If a high-velocity bed underlying a low-velocity overburden is faulted vertically, a refraction profile perpendicular to the strike of the fault can detect the faulting. As illustrated in Fig.39, there is an offset in the

TABLE II

Penetration depth of a curved ray trajectory in a medium for linear increase of velocity* with depth

Shot-detector distance $X(m)$	Maximum depth $Z(m)$	Penetration Z/X (%)
1000	25	2·5
2000	90	4·5
5000	600	12
8000	1450	18
10,000	2130	21
15,000	4105	27
20,000	6300	31·5

*$V_0 = 2400$ m/s; $k = 0·5$ m/s per metre.

travel-time segment corresponding to the high-velocity layer V_2. The travel-time jump ΔT is positive for the down-throw side of the fault. The throw of the fault Z_t is related to ΔT and can be calculated using the following formula:

$$Z_t = \Delta T \, V_2 V_1 \, / \, \sqrt{(V_2{}^2 - V_1{}^2)} \qquad (2.28)$$

The presence of basement faulting is detected more directly and surely in the refraction technique than in the reflection technique.

In the early days of seismic prospecting, the refraction technique was extensively used for locating shallow salt domes. The

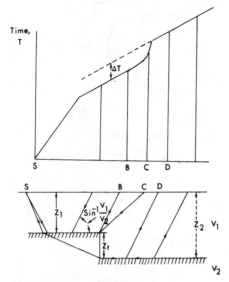

Fig.39. Detection of a fault by refraction shooting. The travel-time jump ΔT is a measure of the throw of the fault. (After Dobrin, 1960.)

shot–detector arrangement (Fig.40) used was much different than that normally used in profile shooting. Detectors were set up at intervals along a sector or a "fan" at a distance of 5–10 km from the shot point. The seismic wave velocity is much greater in salt than in sediments. The passage of a seismic wave through a salt dome causes an appreciable shortening of the travel time, resulting in a "time lead" at some detectors. A second "fan", approximately at right angles to the first fan, enables the position and extent of the salt dome to be outlined. Literally thousands of such fans were shot in the Gulf Coast of Texas and Louisiana during the 1930's in order to locate salt domes.

In completely new areas the fan-shooting method is still valuable, as it covers a wide area with a comparatively limited amount of shooting. In principle, the method could also be used to locate large anticlines or buried valleys provided the time leads (or time lags) caused by the structure are apprecia-

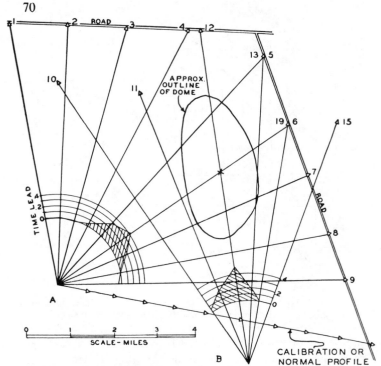

Fig.40.Fan-shooting method for locating salt domes. Hatched areas show the magnitudes of time "leads" corresponding to a dome in the approximate position indicated. (After Nettleton, 1940.)

ble. Griffiths and King (1965) have shown how the method could be applied to trace the course of a buried valley. Scheller (1970) has given an example of the use of fan shooting for detecting buried landslide debris in the Tomba area of Switzerland. An example of the application of fan shooting to investigate crustal structure in Sweden is given by Båth (1971).

Applicability and limitations of the refraction method. Refraction surveying has some important advantages over reflection work. In a virgin area, where no information is available on the subsurface geology, the refraction method is valuable for reconnaissance. In contrast to the reflection surveys, which in the absence of velocity data can give only the geometry of the subsurface formations, refraction surveys yield data on the seismic velocities and geometry of the formations. The additional information on velocities is often of great use when correlating and identifying various formations which are being mapped. For this reason, almost all deep-crustal studies are made with refraction soundings (see section *Crustal seismology*, p.74). For shallow-depth investigations (e.g. civil engineering projects for bedrock depth studies) the refraction method also has the advantage of speed and economy.

On the other hand, the refraction method has some grave disadvantages, too, in comparison with reflection work. First, the method is inherently blind to detect a low-velocity layer (V_2) sandwiched between an upper layer (V_1) and a lower layer (V_3) of higher velocity ($V_3 > V_1 > V_2$). There appears to be no way of determining V_2, nor its thickness, by the refraction method. Fig.41 illustrates this. The effect of the existence of an intermediate low-velocity layer would result in an overestimate of the depth for the underlying refractor, the overestimate being dependent on the thickness of the low-velocity layer and on the velocity contrast involved.

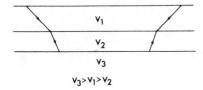

$V_3 > V_1 > V_2$

Fig.41. Low-velocity layer V_2 sandwiched between two high-velocity layers. The refraction method cannot determine V_2.

Another limitation of the refraction method is illustrated in Fig. 42. If one of the layers is thin in comparison with the depth, the refracted wave from it may never reach the surface as a first arrival. For example, in Fig. 42 the first layer (V_1) shows up quite well, as well as the third layer (V_3), but the relatively thin intermediate layer (V_2) never appears as a first

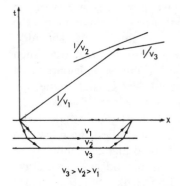

Fig.42.Intermediate-velocity layer V_2 is too thin to appear on the first arrivals of the refraction seismogram, but may give usable secondary signals if the noise is not too high. (After Dix, 1966.)

arrival, and so has to be selected from the background of different energy which usually makes its identification very difficult. The error due to such a "hidden layer" (blind zone) makes the computed depth too shallow, as the overburden is assumed to be less than it actually is. The problem of a hidden layer has been discussed in detail by Kaila and Narain (1970).

In addition, the refraction method is unsuitable for "pin-pointing" a detailed structure, and, at best, yields a simple-layered model which may be a great simplification of a more complicated structure. The method also assumes that there is no lateral variation of the velocity within the layers.

CRUSTAL SEISMOLOGY

The term crustal seismology is used in a restricted sense for the application of seismic studies to crustal structures. Earth-quake seismology laid the foundation for the study of the crust. As a method of investigating the thickness and structure of the crust, it has now largely been superseded by the methods of "explosion seismology", namely the refraction and reflection studies using artificial explosions. The advantage of using artificial chemical (or nuclear) explosions rather than earth-quakes is that the time and position of the shot are accurately known. Also, shots can be fired when required, whereas one has to wait for earthquakes of suitable intensity to occur.

We shall discuss in brief the contribution of seismology to the investigations of the crust.

Modern views about the crust

The crust of the earth is nowadays viewed from two different angles. Seismologically speaking, the crust is the part of the solid earth above the Moho discontinuity, which separates it from the mantle. This discontinuity is marked by a rapid increase in P-wave velocity to about 8 km/s. The worldwide existence of this discontinuity has been established by seismic refraction and reflection studies, though in tectonically unstable regions it often becomes vague and ceases to give clear refractions. The crustal thickness above the Moho ranges widely from about 6 km under the ocean floors to perhaps 70 km under the Andes Mountains.

The other concept of the crust is based mainly on rigidity and isostasy considerations (see p.119). The earth's outermost layer consists of a relatively strong and rigid material. This rigid layer, called the lithosphere, is thought to extend down to an underlying weaker and warmer substratum, called the asthenosphere*. The lower boundary of the rigid crustal layer (lithosphere) may well extend deep into the mantle so that this boundary in general may not coincide with the Moho. Also the change in mechanical properties (rigidity) is now believed to occur at about the top of the low-velocity layer (see p.25). It is still not clear whether the Moho marks a change in composition or in crystal structure. From these considerations, and, in general, as information about the crust and upper mantle has accumulated, the impor-

---

* The asthenosphere is characterized by a low seismic velocity observed at a variable depth in the upper mantle (see Fig. 14).

tance of the Moho has decreased. It is probable that some of the other diffuse boundaries play a more important role in global tectonics than the Moho.

Structure of the continental crust

Early seismological studies by Conrad (1925) suggested that the crust itself is two-layered, and following this suggestion some seismologists designated the upper part of the crust (P velocity 5·6–5·9 km/s) as the "granitic" layer, and the lower crust (P velocity ∼6·5 km/s or more) as the "intermediate layer". This seismic model fitted well with the then popular concept of "sialic" and "simatic" composition of the upper and lower crust.

Modern work from explosion seismology has changed, in detail, the velocity structure, but has not contradicted the basic hypothesis that the crust varies, from top to bottom, between an acidic and an intermediate or basic composition. The important conclusion has been that the two-fold subdivision is not found everywhere. Layering of the crust, or the lack of it, is a region phenomenon, and to some extent it is also a function of the interpreter's methods and ideas as emphasized by Båth (1961).

Most extensive studies of the crustal structure by explosion seismology have been made in the U.S.A., Russia, and Germany. The important results of these studies can be found in special volumes edited by Hart (1969) and Mueller (1974). It is possible here only to summarize some general conclusions and to present a few examples.

The Moho is usually found between depths of 20 and 50 km but regionally it may be deeper, for instance, beneath young fold mountain ranges. In a few special regions, a large number of long-range refraction soundings have been made to study regional variations in the depth to the Moho. Such refraction profiles are typically 200–300 km long and require explosive charges amounting to several tons.

An important example is shown in Fig.43 from studies

(Pakiser, 1963) made in the western part of the U.S.A. It is apparent that the thickening of the crust occurs beneath the mountain ranges and that each of the main geological provinces is associated with a characteristic crustal thickness.

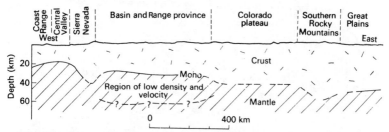

Fig.43. Variations in the crustal thickness from California to Colorado based on refraction surveys. (After Pakiser, 1963.)

Other examples of regional variation in crustal thickness are provided by "deep seismic sounding" (DSS) studies made in the western part of Europe (Closs, 1969) and the U.S.S.R. A summary of the crustal sections which have been obtained in various tectonic zones of the U.S.S.R. has been given by Kosminskaya et al. (1969) and is shown in Fig.44. The regional variations in crustal thickness in many cases may be closely related to the isostatic compensation of the topography. This point is further elaborated in the section *Isostasy and crustal thickness* (p.122).

Recently, attempts have been made, particularly by German seismologists, to study the late reflection phases that occur in records shot during commercial prospecting. Statistical distribution of the consistently occurring late phases suggests reflection times of 4–11 seconds for three discontinuities in some (but not all) of the regions studied (Liebscher, 1964). In Fig.45 the reflections occurring between 10 and 11 seconds (after the shot) are interpreted as being due to the Moho discontinuity, between 7 and 8 seconds as due to the Conrad discontinuity, and those occurring at about 4 seconds are attributed to a new discontinuity named the Förtsch disconti-

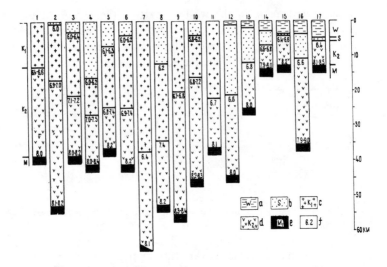

Fig.44.Models of the earth's crust for different tectonic zones of the U.S.S.R. on the basis of deep seismic sounding data. $a=$water, $b=$ unconsolidated crust (sediments), $c=$consolidated crust, upper layers, $d=$consolidated crust, lower layers, $e=$upper mantle, $f=$boundary velocity in km/s. (After Kosminskaya et al., 1969.)

nuity. This discontinuity has also been observed in some refraction studies made by Mueller and Landisman (1966) who consider it to mark the base of a low-velocity channel in the upper crust.

Until about 1960 explosion seismologists were much more sceptical about the widespread existence of the Conrad discontinuity. Later studies using elaborate detecting techniques (close spreads of seismometers along long refraction lines) in the U.S.A. appear to suggest the existence of a lower crustal layer beneath several (but not all) of the structural provinces studied. In any case, there is little support for the idea of a widespread, reasonably homogeneous, lower crustal layer. Thus present-day ideas on the layering of the continental crust are much less definite than those of 30 years ago.

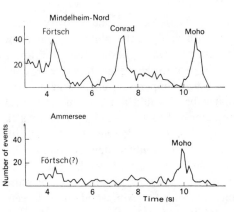

Fig.45.Frequency polygons showing the number of events observed as a function of two-way travel-time (in seconds) in normal-incidence seismic reflection surveys in South Germany. The peaks are interpreted as major discontinuities within the crust (Förtsch and Conrad) and at the base (Moho). (Based on Liebscher, 1964; redrawn from Bott, 1971.)

A brief discussion of the composition of the continental crust would be in order, particularly in the light of observed velocities from the studies of explosion seismology. These studies typically yield P velocities of 5·9–6·2 km/s for the upper crust, which agree with the experimentally determined values for granites at 1 kbar pressure (or 10^8 N/m^2 in SI-units), suggesting a granitic composition for the upper crust. However, this is inconsistent with the average crustal density and chemical composition estimates (Harris, 1971) which suggest a more basic composition than granite. According to Bott (1971, p.67), this discrepancy may be explained by the relative abundance of slates and greywackes in basement rocks; these rock types possess higher densities (2·7 g/cm^3) but have distinctly low P velocities.

The lower crust layer, in areas where it can be distinguished from the upper crust, is characterized by a P velocity of 6·5 km/s or more. Presumably the lower crust has a density of about 3 g/cm^3 compared with an average density of about 2·8

g/cm^3 for the upper crust. These density and velocity values appear to be consistent with gabbro, the plutonic equivalent of basalt, and until recently it was believed, that the lower crust is formed of basalt or gabbro. However, this is now considered doubtful. The pressure–temperature environment in the lower crust is such that gabbro would not exist in its natural form, but would be transformed to a dense high-pressure rock called eclogite. The experimentally determined P velocity for eclogite is greater than 8 km/s and the density is about 3·5 g/cm^3—both too high for the lower crust. Recently it has been suggested that the lower crust consists of some high-pressure modifications of granite and diorite (Ringwood and Green, 1966). The arguments are petrological rather than geophysical.

Structure of the oceanic crust

Information about the oceanic crust has been obtained by geophysical measurements to an even greater extent than for the continental crust. With the aid of earthquake surface-wave studies, it became evident that the oceanic crust was thinner than the continental crust. But it was not until about 1950 that seismic refraction measurements and improved surface-wave analyses firmly established the thickness of oceanic crust at 6 km for all large oceanic areas investigated. The only exceptional areas showing anomalous crustal thickness are volcanic islands, mid-ocean ridges and trenches. Details of the marine seismic techniques together with a compilation of most of the early results can be found in Hill (1957).

One of the important results of the explosion seismology work was the clear demonstration that layering of the oceanic crust is fundamentally different from that of the continental crust in that the granitic layer is completely missing in the oceanic crust. This difference has become all the more significant in the light of the recent discovery that most of the oceanic crust has been formed during the last 200 m.y. (see section *Magnetic record of sea-floor spreading*, p.358), in contrast to the continental crust which was formed as early as 3000 m.y.

The results of refraction measurements show an unexpected uniformity of the oceanic crustal structure. The crust beneath the ocean floor consists of three layers above the Moho. Layer 1, which consists of unconsolidated sediments (P velocity 1·6–3 km/s) is on the average less than 0·5 km in thickness. It is locally absent on the crests of the mid-ocean ridges, but at the other extreme its thickness reaches over 3 km in oceanic troughs. Layer 2 is also variable in thickness (1·2–2·4 km) and velocity (4–6 km/s). Layer 3, which is the main oceanic layer, is the most uniform in thickness (4·5–5 km) and velocity (6·5–7 km/s). Table III summarizes the structure in normal oceans (apart from mid-ocean ridges and continental margins).

A generalized structure section from the North American continental shelf to the Mid-Atlantic Ridge as determined by refraction measurements (Ewing, 1969) is shown in Fig.46. The average depth of the Moho in the Atlantic is about 12 km (reckoned from the water surface), the most significant deviation from this depth occurring on the flanks of the ridge,

TABLE III

Normal structure of the oceanic crust (after Raitt, 1963)

Layer	Thickness (km)	Velocity (V_p in km/s)	Nature
Seawater	4·5	1·5	seawater
Layer 1	0·45	2	unconsolidated sediments
Layer 2	1·7	5·1	basalt ? (or consolidated sediments)
Layer 3	4·8	6·7	serpentine ? (or basalt)
....		Moho	....
Upper mantle		8·1	ultrabasic rock

where mantle velocities have been recorded at depths of 9–10 km. Also characteristic of the ridge is the apparent disappearance of the Moho in the crestal zone, the velocities recorded in this anomalous zone being intermediate between that of layer 3 and the upper mantle. The anomalous structure associated with the ridge crest is possibly related to upwelling convection currents and to the sea-floor tectonics discussed later in Chapter 8.

As to the probable composition of layer 2 and layer 3, seismic velocities alone cannot yield the desired information. The commonly observed range of P velocities of layer 2 (4·4 – 5·6 km/s) could be caused either by weathered lavas or by consolidated sediments. From the evidence of dredged samples, layer 2 is commonly believed to be composed of basaltic material (lavas or intrusions?). Probably the strongest support for this belief has come from the evidence of oceanic magnetic anomalies, which conclusively show that highly magnetic rocks form a substantial part of layer 2. Layer 3 forms the main thickness of the oceanic crust. From P-velocity considerations (6·5–

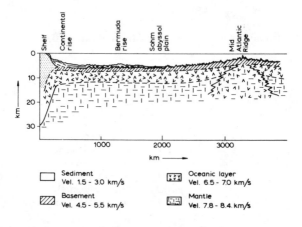

Fig.46.Generalized structure section from the North American continental shelf to the Mid-Atlantic Ridge. (After Ewing, 1969.)

7 km/s), it can be interpreted as a consolidated basic igneous rock such as gabbro. The discovery of metamorphosed basalts and serpentinized peridotite in dredge hauls from the Mid-Atlantic Ridge has opened further possibilities for speculation. According to Hess (1962), the oceanic Moho represents the boundary between the normal upper mantle and its serpentinized alteration product (see Fig. 180). Cann (1968) considers layer 3 to be formed by the metamorphism of the basaltic crust to amphibolite in the crestal zone of the ridge where the temperature is high. Until the basic question about the composition of layer 3 is resolved various speculations will continue to remain.

Use of surface waves in crustal studies

The practical use of surface waves in the study of crustal structure has only eventuated in recent decades. We noted earlier (p.14) that surface waves exhibit dispersion whose characteristics depend on the layering involved. The advances in instrumentation and methods of data analysis have made the dispersion studies a powerful tool for investigating the crustal structure. The technique employed is to read the wave periods on a seismogram as time advances and, knowing the epicentre and origin time of the shock, to convert the information into a velocity–period curve, or a dispersion curve. This curve is then compared with theoretical curves based on models of the earth's crustal structure. The best fitting model is accepted. The theoretical computation of model dispersion curves is an advanced and complicated mathematical operation and will not be treated here. For a fuller discussion of the subject, the reader is referred to Kovach (1966).

An example of crustal studies of the Greenland shield by surface waves is given in Gregersen (1970). Fig.47 shows a computed crust model for the Greenland shield based on dispersion data from usable earthquakes in the period January 1964–April 1967. In this model, the ice cap is assumed to be a homogeneous layer with a mean thickness of 2·5 km. The P

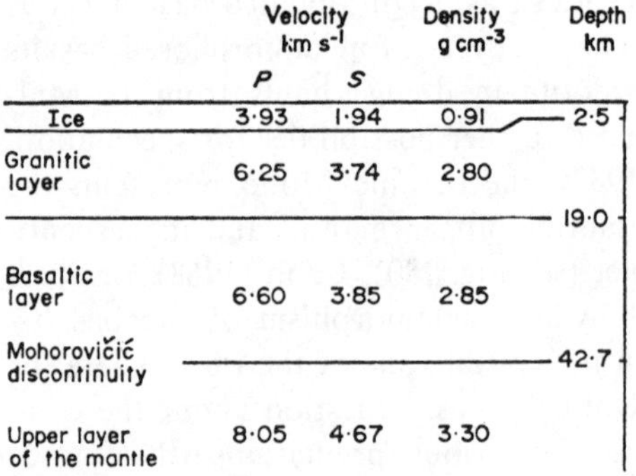

	Velocity km s^{-1}		Density g cm^{-3}	Depth km
	P	S		
Ice	3·93	1·94	0·91	— 2·5
Granitic layer	6·25	3·74	2·80	
				— 19·0
Basaltic layer	6·60	3·85	2·85	
Mohorovičić discontinuity				— 42·7
Upper layer of the mantle	8·05	4·67	3·30	

Fig.47.Crust model G-62 for the Greenland shield derived from dispersion studies of surface waves. (After Gregersen, 1970.)

velocity of 3·93 km/s, the S velocity of 1·94 km/s, and the density of 0·91 g/cm³ for ice were taken from the deep borehole data of Camp Century, in north Greenland. The velocities agree well with those found by Holtzscherer (1954) using the seismic refraction method. The Greenland shield model has a crust thickness of about 40 km and is characterized by high granitic P velocities of 6·2–6·3 km/s.

A substantial amount of our knowledge of the large-scale structure of the earth's crust and upper mantle, and the difference between continental and oceanic crusts, has been derived from the study of surface-wave dispersion. The simplest classification of crustal types into continental and oceanic types remains fundamentally correct. However, a more refined classification that has emerged from surface-wave studies leads to a division of the earth's crust into the following types:

(1) A continental crust overlying a stable mantle, i.e. shield and mid-continent types.

(2) A continental crust overlying an unstable mantle, i.e. basin-range, Alpine, and great island-arc types.

(3) An oceanic crust overlying a stable mantle, i.e. deep-ocean basin types.

(4) An oceanic crust overlying an unstable mantle, i.e.

TABLE IV

Classification of crustal types (after Brune, 1969)

Crustal type	Tectonic characteristic	Crustal thickness (km)	P_n*1 (km/s)	Heat flow (HFU)*2	Bouguer anomaly (mgal)	Geologic features
Continental crust overlying stable mantle						
(A) Shield	very stable	35	8·3	0·7–0·9	−10 to −30	little or no sediment, exposed batholithic rocks of Precambrian age. Moderate thicknesses of post-Precambrian sediments.
(B) Midcontinent	stable	38	8·2	0·8–1·2	−10 to −40	
Continental crust overlying unstable mantle						
(C) Basin-range	very unstable	30	7·8	1·7–2·5	−200 to −250	recent normal faulting, volcanism, and intrusion; high mean elevation.
(D) Alpine	very unstable	55	8·0	variable 0·7–2·0	−200 to −300	rapid recent uplift, relatively recent intrusion; high mean elevation.
(E) Island arc	very unstable	30	7·4–7·8	variable 0·7–4·0	−50 to +100	high volcanism, intense folding and faulting.
(F) Plateau	not adequately studied					

(contd.)

TABLE IV (contd.)

Crustal type	Tectonic characteristic	Crustal thickness (km)	P_n[*1] (km/s)	Heat flow (HFU)[*2]	Bouguer anomaly (mgal)	Geologic features
Oceanic crust overlying stable mantle						
(G) Ocean basin	very stable	11	8·1–8·2	1·3	+250 to +350	very thin sediments overlying basalts, linear magnetic anomalies, no thick Palaeozoic sediments.
Oceanic crust overlying unstable mantle						
(H) Ocean ridge	unstable	10	7·4–7·6	high and variable 1·0–8·0	+200 to +250	active basaltic volcanism. little or no sediment.
(I) Ocean trench	not adequately studied					

[*1] Mantle velocity corresponding to the Moho discontinuity.
[*2] To convert the heat flow values to SI-units (mW/m²) multiply by 41.9.

mid-ocean ridge and deep-ocean trench types.

Some important characteristics of these crustal types are summarized in Table IV. It will be noted that tectonically unstable crustal types (C, D, E, and H) are characterized by a high heat flow, low P_n velocities and a more pronounced low-velocity channel.

An excellent account of the use of surface-wave dispersion in quantitative studies of crust and mantle structure is given in Brune (1969).

CHAPTER 3
Gravity Methods

INTRODUCTION

Since the celebrated earth–apple relationship first discovered by Newton, the mutual attraction between all masses has been recognized as a universal phenomenon. This phenomenon accounts for the familiar fact that a body released near the earth will fall with increasing velocity. The rate of increase of velocity is called the gravitational acceleration, g, or, simply, the gravity which Galileo proved to be the same for all bodies at a given location on the earth.

If the earth were a perfect sphere of uniform concentric shell structure, the force of attraction on a body lying on its surface would be the same everywhere, and the gravity would have a single constant value. In fact, our earth is non-uniform, non-spherical, and rotating, and all these facts contribute to variations in the gravity over its surface. However, these variations are relatively small, and their accurate measurement is possible only with the aid of extremely sensitive instruments.

The measurements and analyses of the variation in gravity over the earth's surface have become powerful tools in the investigation of the earth. The variations in gravity which are related to the departure of the earth from a spherical form are of particular interest in geodesy, the science of the earth's shape. On the other hand, variations in gravity which reflect lateral contrasts in the density of subsurface rocks are of prime interest to the geologist for extracting information about structures at various depths. This information is always subject to a certain ambiguity inherent in the interpretation of gravity field data, although in practice the ambiguity is reduced to a considerable extent by additional geological information available from other independent sources.

As a geophysical tool, the gravity method has a great deal in common with the magnetic method. Both seek anomalies caused by changes in physical properties of subsurface rocks. Also, both require fundamentally similar interpretation techniques. The usefulness and applications of the two methods, however, vary considerably depending on the relative prominence of the physical property contrast involved in various geological situations.

We shall first outline the physical principles which are basic to a proper understanding of the gravity measurements and their interpretation. We shall also discuss in brief the role of gravity measurements in studying the isostasy and crustal structure of the earth. This will be followed by several examples of geological problems concerning regional and local structures in which gravity surveys have been of great assistance.

FUNDAMENTAL PRINCIPLES

Newton's law

The force of attraction, F, between two particles of masses m_1 and m_2, separated by a distance, r, is given by:

$$F = G \; \frac{m_1 m_2}{r^2} \qquad (3.1)$$

where G is the universal gravitation constant. The numerical value of G was not determined in Newton's life time, and was first measured in the laboratory by Cavendish in 1798. The currently accepted value for G is 6.67×10^{-11} m^3/kg s^2 in SI-units. In the c.g.s. system of units G is 6.67×10^{-8} (cm^3/g s^2), which is numerically equal to the force in dynes that will be exerted between two masses of 1 g each with centres 1 cm apart. In SI-units, force F is expressed in newton ($1 N = 10^5$ dynes).

Gravitational acceleration

The force exerted on a body at the earth's surface is due to the attraction of the earth. When the effects of rotation and non-uniformity of the shape and density of the earth are neglected, the force exerted on the body of mass m is:

$$F = \frac{GM_E\,m}{R^2} \tag{3.2}$$

where M_E is the earth's mass and R is its mean radius.

The force acting on the body is also given by Newton's second law of motion; i.e., $F = mg$, where g is the acceleration that would be caused by the gravitational pull of the earth if the body were allowed to fall freely. Thus the gravitational acceleration g (henceforth referred to only as gravity) may be considered as the force exerted by the earth on a unit mass, and it can be expressed as:

$$g = F/m = \frac{GM_E}{R^2} \tag{3.3}$$

In the c.g.s. system the unit for g is cm/s^2. Among geophysicists this unit is referred to as gal (in honour of Galileo). The value of g has a worldwide average of about 980 gal or $9\cdot8$ m/s² in SI-units on the earth's surface, with a total range of variation from the equator to the poles of about 5 gal, or approximately $0\cdot5\%$ of g. The practical unit commonly used in geophysics for the measurement of g is the milligal (mgal). A tenth of a milligal is called a gravity unit (g.u.) which is used as a subunit of gravity acceleration in SI (1 g.u. = 1 μm/s²). The unit mgal has been retained in this book, because of its extremely wide use in geodesy and geophysics.

The earth's mass and density

The mass of the earth can be estimated from the value of the gravity at the surface, after correcting for the small contribution to the gravity caused by rotation. For this purpose the relation given by eq.3.3 is applicable. During the 18th century attempts to measure G were often considered to be experiments to determine the mass of the earth, or equivalently, the mean density of the earth. The results of Cavendish showed that the mean density (ρ_E) was about 5·4 g/cm³ (or 5·4 × 10³ kg/m³ in SI-units), and this was one of the first proofs that the earth's interior consists of material considerably denser than surface rocks. Using more precise values for g and G, Jeffreys (1970) obtained the following values for the earth's mass and mean density: $M_E = 5\cdot977 \times 10^{24}$ kg; $\rho_E = 5\cdot517$ g/cm³. The SI-unit for density is kg/m³; however, the sub-unit g/cm³ has been retained in this book because of its widespread use in the geophysical literature.

The moment of inertia, I, of the earth also provides some information about the mass distribution within the earth. I is not directly measurable, but it can be determined from the earth's dimensions, mass, gravity, and angular velocity. The value for I as determined from observations of satellite orbits, is $0\cdot331 M_E R^2$. If the earth were a uniform sphere, its moment of inertia would be $0\cdot4 M_E R^2$. Thus the earth has a lesser moment of inertia than a uniform sphere. This means that the mass is concentrated toward the centre of the earth.

Knowledge of the mean density and moment of inertia is important in studying the internal density distribution within the earth, because any acceptable model of density must satisfy these observations (see section *Density and other parameters in the earth*, p.28).

Gravitational potential

The gravitational field around a point mass is a vector field as shown in Fig.48. This field is given by the force of attraction exerted on a unit mass and the field vectors are directed

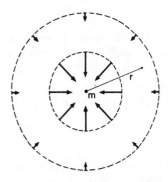

Fig.48.Sectional illustration of the gravitational field around a point mass (*m*). Field lines are shown by arrows. Equipotential surfaces are shown by dashed lines.

towards the attracting mass.

The potential at a given point in a gravitational field is defined as the work done by gravity to move a unit mass from an infinite distance to the point in question. If the unit mass is brought from infinity, it can be shown that the energy necessary (or, equivalently, the work done) to move it to a final position at a distance, r, from an attracting source of mass m is Gm/r. In other words, the gravitational potential U, often referred to as the Newtonian potential, is:

$$U = Gm/r \tag{3.4}$$

A gravitational field can be represented by surfaces over which the potential is constant, known as equipotential surfaces. The force vectors are normal to these surfaces, so that there is no component of force along them. The surface of a mass of fluid in a gravitational field coincides with an equipotential surface, and for this reason the potential is a quantity of great importance in the study of the shape of the sea-level surface of the earth. The potential function, U, also plays an important part in the theory of attraction; the negative derivative of U gives the gravity attraction in the corresponding direction (e.g. $g_z = -\partial U/\partial z$).

Some knowledge of the potential field theory is essential to appreciate the methods of analysing and interpreting the potential field data such as gravity, magnetic, and electric. An excellent treatment of the subject, relevant to the needs of the geophysicists, will be found in Grant and West (1965, chapter 8).

The normal spheroid and the geoid

The earth is not a perfect sphere. Its shape approximates very closely to that of a perfect fluid for which a balance is maintained between the gravitational forces tending to make it spherical, and the centrifugal forces of rotation tending to flatten it. As a result of this balance the equatorial radius, R_E, is about 21 km greater than the polar radius, R_p. Because of this flattening, g is distinctly less at the equator than at the poles, the actual difference being about 5200 mgals (or $5 \cdot 2 \times 10^{-2}$ m/s²).

Since the introduction of pendulum measurements, the analysis of gravity data over the earth's surface has played an important role in describing the shape of the earth in terms of an ideal reference surface. Although the most accurate reference surface appears to be a spheroid rather than an ellipsoid of revolution, the latter is more convenient to use, since fewer constants are involved. The variation of gravity with latitude over the surface of an ellipsoidal earth can be expressed in the form:

$$g = g_0 (1 + C_1 \sin^2 \phi - C_2 \sin^2 2\phi) \qquad (3.5)$$

where g_0 is the value of the gravity on the equator and ϕ is the latitude. C_1, C_2 are constants depending on the earth's shape, the numerical values of which are adjusted to give a best fit to the measured variation of the gravity over the earth's surface.

Until recently, the International Gravity Formula (1930)

was the accepted standard for the reduction of gravity data. According to this formula, the normal sea-level gravity at a latitude ϕ is given by:

$$g_\phi = 9\cdot78049 \ (1+0\cdot0052884 \ \sin^2 \phi - 0\cdot0000059 \ \sin^2 2\phi)$$

$$m/s^2 \qquad (3.6)$$

Here $9\cdot78049$ is the derived value of the equatorial sea-level gravity based on the Potsdam standard. The constants, C_1 and C_2, correspond to a flattening of $1/297$ for the spheroid (taken to be an ellipsoid of revolution) with $R_E = 6378\cdot388$ m, and $R_p = 6356\cdot909$ m.

Recent studies on the orbits of artificial satellites have provided more precise values for constants, C_1, C_2, and the following is the revised geodetic standard established by the I.U.G.G.* in 1967:

$$g_\phi = 9\cdot780318 \ (1+0\cdot0053024 \ \sin^2 \phi - 0 \ 0000059 \ \sin^2 2\phi)$$

$$m/s^2 \qquad (3.7)$$

If the earth were a perfect fluid with no lateral variations in density, its surface would correspond to the reference, or so-called, normal spheroid. In using any reference spheroid for practical work, one must relate the equipotential surface to some physical surface on the earth. Over the oceans this surface is the mean sea level, and on land the surface could be related to the water surfaces in narrow sea-level canals if they were extended inland. The level surface so defined is called the "geoid". By definition, the geoid is horizontal everywhere, and the direction of gravity would be normal to this surface.

In the case of the actual earth, the geoid will not generally

*International Union of Geodesy and Geophysics (Geodetic Reference System, 1967). The difference between the 1967 formula and the 1930 formula is given by: $g_{\phi(1967)} - g_{\phi(1930)} = (-172 + 136 \sin^2\phi) \ \mu m/s^2$.

coincide with the reference spheroid. This is because there are local distortions of the geoid caused by the irregularities in density within the earth and by the topographic irregularities of its surface. These irregularities result in localized mass anomalies. The effect of a mass anomaly on the equipotential surface of the gravity field is illustrated in Fig.49. Over a region of mass excess, there is an additional potential, ΔU, which causes the equipotential surface to warp outward.

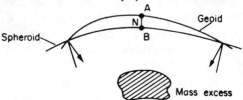

Fig.49.Undulation of the geoid (N) and deflection of the vertical caused by a mass anomaly. (After Garland, 1965.)

For a single mass anomaly in an otherwise uniform earth, the warping N of the geoid is given by:

$$gN = \Delta U \tag{3.8}$$

where g is the mean value of the gravity between A and B. On either side of the region of mass excess, the plumb line is deflected inward. Over a region of mass deficiency the effect would be opposite.

Local undulations of the geoid provide valuable information on the subsurface structures of the earth. On the other hand, large-scale depressions and elevations of the geoid surface, as determined by satellite observations, reflect deep-seated density anomalies whose source must lie within the mantle (see Fig.69).

Rock densities

If the earth were made up of a series of shells of horizontally uniform density, there would be no horizontal variation

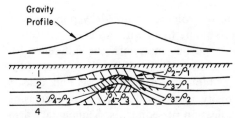

Fig.50.Schematic section showing lateral density contrasts resulting from a structural uplift. ρ_1, ρ_2, ρ_3, and ρ_4 are the densities of four flat-lying layers. Horizontal variation in gravity is caused by lateral variations in density. (After Nettleton, 1971.)

in gravity no matter what the vertical variation in density might be. Any geologic condition that results in a horizontal variation in density will cause a horizontal variation in gravity or a gravity "anomaly". This is illustrated in Fig.50 where the flat-lying layers are disturbed by a structural uplift. For the interpretation of gravity anomalies, it is therefore necessary to estimate the differences in densities (density contrasts) between the subsurface rocks before one can postulate their structure.

The *bulk* density of a rock is controlled by three factors: the grain density of minerals forming the rock mass, the porosity, and the fluid in the pore space.

For most sedimentary rocks the density is controlled, to a very large extent, by their porosity. It is, therefore, necessary to distinguish between the "bulk" density and the "grain" density, ρ_g. A further distinction is usually made between dry and wet bulk density. Dry bulk density, ρ_d, refers to specimens in a completely moisture-free state, whereas wet bulk density, ρ_w, implies that the specimen is fully saturated with water. The interrelationship between ρ_g, ρ_d, and ρ_w is given by the following formulae:

$$\rho_d = \rho_g \ (1 - P) \qquad (3.9)$$

$$\rho_w = \rho_d + Pa \qquad (3.10)$$

where P is the fraction of volume occupied by pores, and a is the fraction of pore volume filled with water.

In the case of highly porous materials, the difference between ρ_d and ρ_w may be as much as 30%. For most rocks below the ground-water table it is safe to use wet bulk density values. As a rule porosity tends to decrease with increasing consolidation and lithification of sediments. Sedimentary densities thus increase from low values $(2\,g/cm^3$ or less), in the case of younger unconsolidated deposits, to values approaching mineral grain densities in older formations. Shales and clays in particular exhibit a remarkably linear variation in density with depth of burial. Table V shows the density values commonly encountered in typical rocks and rock-forming minerals.

While the extreme range of density variations in sedimentary and igneous rocks may appear to be rather large, the actual range of density contrasts likely to be encountered in a geologic structure is considerably less. As a matter of experience, a reasonable figure to keep in mind as a density contrast, commonly involved in geologic structures, is approximately $0 \cdot 25$ g/cm³, and often such a general figure can be used as a first estimate when no definite density information is available. Alternatively, the Nafe-Drake relationship between the seismic P-wave velocity and density (see Fig.9) may be used to estimate the density of shallow rocks from the knowledge of P velocity. In any case, a quantitative interpretation of gravity anomaly data in terms of geologic structure demands a reasonably reliable estimate of the density contrasts involved.

Density is normally measured by weighing a rock sample first in air, then in water. The weight in air divided by the loss of weight in water gives the density. Precautions must be taken in the case of a porous sample (by coating it with a thin layer of melted paraffin wax) to prevent water from entering the pores. Often it is advisable to determine both dry (after oven-drying the sample) and wet (after impregnating the sample with water) bulk densities. The density of the rock in situ may be assumed nearer to ρ_d or ρ_w depending on local conditions.

TABLE V

Densities of rocks and minerals

Rock type or mineral	Density (wet) (g/cm³)*	Remarks
Sand	1·6 −2	data taken mostly from
Moraine	1·5 2	a compilation made by
Sandstones (Mesozoic)	2·15 2·4	Parasnis (1971)
Sandstones (Palaeozoic and older)	2·35−2·65	
Quartzite	2·60 2·70	
Limestone (compact)	2·5 −2·75	
Shales (younger)	2·1 −2·6 (2·4)	
Shales (older)	2·65−2·75 (2·7)	
Gneiss	2·6 2·9 (2·7)	
Basalt	2·7 −3·3 (2·98)	
Diabase	2·8 −3·1 (2·96)	
Serpentinite	2·5 2·7 (2·6)	
Gypsum	2·3	
Anhydrite	2·9	
Rocksalt	2·1 2·4 (2·2)	
Zincblende	4·0	
Chromite	4·5 −4·8	
Pyrite	4·9 5·2	
Haematite	5·1	
Magnetite	4·9 5·2 (5·1)	
Galena	7·4 −7·6	
Granite	2·52 2·81 (2·67)	the following data is
Granodiorite	2·67 2·79 (2·72)	taken from tables by
Syenite	2·63 2·90 (2·76)	Clark (1966)
Quartzdiorite	2·68 2·96 (2·81)	
Gabbro	2·85 3·12 (2·98)	
Peridotite	3·15 3·28 (3·23)	
Dunite	3·20 3·31 (3·28)	
Eclogite	3·34 3·45 (3·39)	

*To convert to S.I.-units (kg/m³) multiply by 10^3.

It is not always easy to make representative density determinations because of the difficulty of finding fresh, unweathered rock samples. Wherever possible the laboratory determinations should be supplemented by in-situ determinations of
density. Various field methods for the in-situ determination of
density are described later (see section *Field determination
of density*, p.115).

GRAVITY EFFECTS OF SIMPLE MASS SHAPES

The calculations of the gravity effects of a few geometrical
models serve as useful guides for estimating the magnitude and
form of the gravity anomalies that may be expected from geological structures, even though the actual forms of the structures
rarely show a close resemblance to the simple geometrical
shapes. Five simple geometric forms, the sphere. the horizontal
cylinder, the horizontal slab, the vertical sheet, and the vertical
cylinder are often used to approximate a wide range of geological structures. We shall briefly discuss some of the important
characteristics of their gravity effects.

(1) *Sphere*. Not many geological structures (excepting some
salt domes) approximate to a spherical form, yet this simple
model is often used as a first estimate for compact and roughly
equal dimensional bodies whose anomalies are nearly circular
in shape. The gravity attraction of a sphere at an external
point is the same as if the entire mass were concentrated at its
centre. With the notation of Fig.51, the gravity anomaly caused
by a sphere is given by:

$$\Delta g = \frac{4\pi R^3 G \ \Delta\rho}{3z^2} \ \frac{1}{(1 + x^2/z^2)^{3/2}} \tag{3.11}$$

$$= \frac{\Delta g_{max}}{(1 + x^2/z^2)^{3/2}} \tag{3.12}$$

where Δg_{max} occurs at $x=0$. It is clear from the curve show-

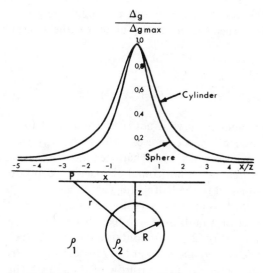

Fig.51.Horizontal variation in gravity due to a sphere and a horizontal cylinder with density contrast $(\Delta\rho = \rho_2 - \rho_1)$. The maximum anomaly occurs at $x=0$.

ing the variation of gravity with horizontal distance, that for a particular value of x/z, Δg approaches a value equal to $\Delta g_{max}/2$. The horizontal distance, over which Δg_{max} falls to half its value, is called the "half-width", $x_{1/2}$, of the anomaly. This parameter is a useful index for estimating the depth. For the sphere the depth to the centre z_c equals $1\cdot305x_{1/2}$. It is, therefore, possible to estimate the depth to the centre of a roughly spherical body directly from the observed anomaly profile without calculations. Moreover, if the density contrast $\Delta\rho$ could be reasonably estimated, the size of the body can also be determined by using eq.3.11.

(2) *Horizontal cylinder.* The gravity attraction of structures which are elongated in the strike direction can be accounted for in terms of the effect of a horizontal line element or that of a long horizontal cylinder. Steep anticlines and buried ridges can be simulated by long horizontal cylinders. Using the same nota-

100

tion of Fig.51, the gravity anomaly due to an infinitely long cylinder (striking perpendicular to the plane of the paper), is:

$$\Delta g = \frac{2\pi G R^2 \, \Delta \rho}{z} \, \frac{1}{(1 + x^2/z^2)} \qquad (3.13)$$

In this case the anomaly curve is not so sharp as that for a sphere at the same depth, since here Δg varies inversely as z (unlike z^2 for a sphere), which is a characteristic of all two-dimensional structures. For a horizontal cylinder the depth to the centre z_c is just equal to the half width $x_{1/2}$.

(3) *Horizontal slab*. This model is very useful for approximating the gravity effect of a fault or a step-like structure. In the case of a simple fault (Fig.52), a horizontal boundary between an upper layer of density ρ_1 and a lower layer of density ρ_2 is displaced vertically by the distance t (throw of the fault). The gravity anomaly of the fault would, therefore, be equivalent to that of a semi-infinite slab of thickness t and density contrast $\Delta \rho$ ($= \rho_2 - \rho_1$), terminated at the fault face ($x=0$).

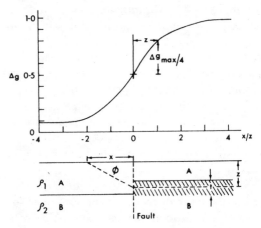

Fig.52.Form of the gravity anomaly across a fault. Δg_{max} is the total change of gravity across the fault. (After Griffiths and King, 1965.)

For a small fault ($t<z$) it can be considered that the mass of the anomalous slab is condensed onto a thin sheet at the median plane of the slab at depth z. The excess mass is, therefore, $\Delta\rho t$ per unit area of the sheet and the resulting gravity effect is $G\,\Delta\rho t$ multiplied by the solid angle subtended by the sheet at the point of observation. For a semi-infinite sheet, using the notation of Fig.52, this solid angle is 2ϕ which is twice the angle measured in the plane of the diagrams. Accordingly the gravity effect becomes:

$$\Delta g = 2Gt\,\Delta\rho\phi = 2Gt\,\Delta\rho\,\left(\frac{\pi}{2} - \tan^{-1}\frac{x}{z}\right)^{*} \quad (3.14)$$

The total change of gravity anomaly across the fault is $2\pi\,Gt\,\Delta\rho$ (which is the gravity attraction due to an infinite slab of thickness t) irrespective of depth z of the slab, and over either side of the fault trace the change is half of the total. The change is also independent of the slope of the lateral face of the slab. Hence product $t\,\Delta\rho$ of a slab can be directly determined from its gravity anomaly. The depth, z, to the median plane of the slab is given by the distance from the fault to the point where the gravity change is one quarter of the total change of anomaly.

The approximation of condensing the mass of the slab onto a thin sheet is surprisingly close even for bodies of considerable thickness such as deep sedimentary basins. For thick slabs, which extend to the surface ($t = 2z$), and which have a horizontal extent of the order of only three times the thickness, Nettleton (1971) gives the practical and easily remembered figure of 100 ft (about 30 m) for the thickness of material with unit density contrast ($\Delta\rho = 1$ g/cm^3) to cause a gravity effect of 1 mgal. The corresponding theoretical value for a slab with infinite horizontal extent (Bouguer slab) is 24 m for 1 mgal.

*For a point of observation on the upthrow side of the fault the half solid angle ϕ would be ($\pi/2 + \tan^{-1} x/z$).

(4) *Vertical sheet*. This model is often used to approximate the gravity effect of tabular bodies (e.g. dikes) elongated in the strike direction. Using the notation of Fig.53 the formula for

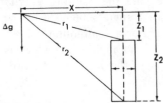

Fig.53.The vertical sheet model, often used to approximate the gravity effect of thin dikes (for explanation see text).

the gravity anomaly of a thin vertical sheet $(t < z_1)$ striking perpendicular to the plane of the paper, is:

$$\Delta g = 2Gt \, \Delta\rho \, \ln (r_2/r_1) \qquad (3.15)$$

The anomaly of a dike increases infinitely with its depth extent, but the increase is logarithmic and very slow.

(5) *Vertical cylinder*. This form is often convenient for approximating the gravity effects of salt domes and volcanic plugs. Using the notation of Fig.54 the expression for the gravity anomaly due to a short cylinder $(t < z)$ is simply:

$$\Delta g = Gt \, \Delta\rho\phi \qquad (3.16)$$

where ϕ is the solid angle subtended by the cylindrical cross-section (in the median plane) at the point of observation. The solid angle, ϕ, can be conveniently determined from a graphical chart designed by Nettleton (1971, p.60). For long, chimney-like bodies (such as volcanic necks and narrow salt plugs) a simpler model of a long vertical cylinder $(t > z_1 > R)$ can be used. In this case, the gravity effect is approximated by the formula for a vertical line element, and is given by:

$$g = \pi R^2 G \, \Delta\rho \big/ \sqrt{(x^2 + z_1^2)} \qquad (3.17)$$

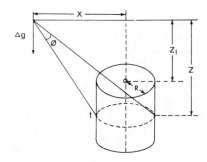

Fig.54.The vertical cylinder model. It is used to approximate the gravity effects of salt domes and volcanic plugs (for explanation see text).

For a vertical line element depth $z_1 = x_{1/2}/\sqrt{3}$ where $x_{1/2}$ is the half-width of the anomaly, as mentioned previously.

GRAVITY MEASUREMENTS AND DATA REDUCTION

Absolute measurements

Measurements of the earth's gravity have been made since the time of Newton. The measurement of g involves only the fundamental dimensions of length and time. However, the absolute determination of g at some point on the earth with a precision of about one part in 1 million is an experiment which requires the greatest care. Almost all the earlier determinations used some form of swinging pendulum whose period is related to g.

Any real body which can be made to swing on a pivot constitutes a physical pendulum, for which the period of oscillation is:

$$T = 2\pi \sqrt{I/mgh} \qquad (3.18)$$

where I is the moment of inertia of the pendulum about the point of support, m its total mass, and h the distance from the point of support to the centre of the mass.

In order to measure g with a precision of one part in 1 million, one would have to know m, I and h to the same degree of precision. Since neither I nor h can be measured with this precision, the above expression is unsuitable for the calculation of g. For every physical pendulum with a given point of support, there is an "equivalent simple pendulum" of length $l\,(= I/mh)$ which has the same period T. If any two points, P and P', can be found on opposite sides of the mass centre so that the periods of oscillation are equal, the distance PP' gives the length of the equivalent simple pendulum, and g can be determined by the simple formula:

$$T = 2\pi \sqrt{l/g} \qquad (3.19)$$

The above principle was first used by Kater in 1818 to measure the absolute value of g at a site in London. Kater's method formed the basis of measurements made at Potsdam by Kuhnen and Furtwängler (1906). These measurements led to the adoption of the Potsdam standard value of $g = 981\cdot274$ gal,* which is known to be in error by several milligals. Later more precise measurements have been made at Washington (1936), Teddington (1940) and Princeton (1965). The observations with a pendulum are extremely tedious and despite the best use of available methods of measuring time and length, the absolute determinations are estimated to be accurate at best within 1– 1·5 mgal (i.e. $10-15\ \mu\text{m/s}^2$).

Another direct approach for measuring g is to time a free-falling body over a known distance. However, it is only recently that timing standards adequate for the precise measurement of short time intervals have become available. A body dropped from rest falls very nearly 5 m in the first second, and taking 5 m to be a practical, reasonably long drop, a time of the order of 1 second must be measured with a precision of one part

---

*The revised Potsdam value (Geodetic Reference System, 1967) is $981\cdot260$ gal ($= 9\cdot81260$ m/s^2).

in 1 million. The method has been highly refined at various laboratories, and good determinations give results accurate to about one part in 5 million (i.e. $\sim 2\mu m/s^2$ or $0\cdot2$ mgal). For technical details of the free-fall experiment the interested reader may refer to Cook (1973).

Relative measurements

Gravity differences are far easier to determine than absolute values. Referring to eq. 3.19, it is easy to see that if the same pendulum is swung under identical conditions at two places, with gravity g_1 and g_2, the corresponding periods T_1 and T_2 will be related by the equation:

$$g_1/g_2 = T_2^2/T_1^2 \qquad (3.20)$$

Thus, for small changes in the period, the difference in gravity can be calculated by the equivalent equation:

$$g_2 - g_1 = \Delta g = \frac{- 2g_1 (T_2 - T_1)}{T_1} \qquad (3.21)$$

Using eq. 3.21, the relative gravity at two places can be determined more accurately than the gravity itself, since the accuracy of Δg in this case primarily depends on that of ΔT (i.e. $T_2 - T_1$). With chronometers as standard timing devices, and by a careful design of the experiment under identical conditions, the accuracy of relative measurements by pendulum should be possible to about $0\cdot1$ mgal. Relative measurements made between the principal sites of absolute determinations indicate the discrepancies between the absolute values themselves, and it is now apparent that the 1930 Potsdam value is approximately 14 mgal too high. Since the development of another class of instruments, which are quicker and more accurate for making relative measurements, pendulums have assumed a secondary role of providing the calibration standards for other gravity measuring instruments.

Gravimeters

The demands of exploration geophysics (chiefly for petro-leum) have led to the development of very precise, portable and fast operating instruments for measuring small variations in gravity. Such instruments (called "gravimeters") are capable of measuring Δg to a precision of 0·01 mgal. The great increase, in recent years, in the number of gravity stations on the earth is solely due to their development.

In principle a gravimeter is an extremely sensitive weighing device whose responsive element is essentially a spring carrying a fixed mass. Since the displacements of the mass produced by small changes in gravity are very minute (of the order of 10^{-10} m), the major designing problem of the gravimeter is the achievement of the necessary sensitivity.

During the last 30 years several ingenious gravimeter de-signs have been developed. One of the most well-known and widely used is the Worden gravimeter (Fig.55). Here the astatic system is designed so that, when its sensitive element is dis-placed by a change in gravity, other forces tending to increase the displacement come into play. The gravity change is measured in terms of the restoring force necessary to return the sensitive element to a standard null position. The essential elements of the Worden meter are all of quartz and are very light in weight. Temperature effects on the system are reduced by a built-in self-compensating device and, in addition, the entire system is kept in a small sealed thermos flask. The total weight of the instrument, including its carrying case, is about 5 kg. The in-strument can be read to 0·01 mgal and the model used in explora-tion work has a dial range of about 100 mgal. A special model designed for geodetic work has a range of about 5500 mgal, so that large differences in g can be measured, although with lower precision. Instrumental designs of the Worden and some other makes can be found in Dobrin (1960) and Jung (1961).

The reading of a gravimeter at any point depends on where the dial scale is set and bears no relation to the absolute value of gravity at that point. If the gravimeter is carried around for

Fig.55.Worden gravimeter in use in Greenland.

a few hours, or even left untouched at one place and then read again later at that place, a change in the reading will be noticed. If additional readings are taken over a period of hours at the same place and the observed gravity is plotted against time, it will be found that the points tend to fall on a smooth curve. This continual variation of the gravity readings with time is known as "drift" and is caused by the fact that the gravimeter springs are not perfectly elastic but are subject to a slow creep over long periods. The usual method of correcting for the drift is to repeat the measurements at the base station at intervals of 1–2 hours. From such a drift curve the base reading corresponding to the time a particular station was measured can be obtained. This base reading is to be subtracted from the reading at the station to obtain the gravity difference. A sample drift curve of a Worden gravimeter is reproduced in Fig.56.

Difference in scale divisions converted to mgal.

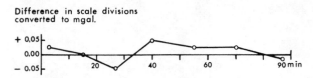

Fig.56. A typical drift curve of a Worden gravimeter. (1 mgal = 10 g.u.)

Since all the readings of gravimeters are in arbitrary scale divisions, calibration is necessary to express them in milligals. The value of the calibration constant for every instrument is supplied by the makers, but as it can change with time due to elastic creep in the springs, intermittent checks on it are usually necessary. For checking the calibration constant the most usual method is to read the gravimeter at two stations between which the gravity difference is accurately known from pendulum measurements. A more convenient but less precise method is to read the instrument at the top and the bottom of a tower, or of a tall building. In this case it can be shown that the gravity difference, Δg, will be $0\cdot3086h$ mgal where h is the elevation difference (in metres) between the two points of observation.

Our knowledge of the earth's gravity field would be severely restricted if gravity measurements were not extended to water-covered areas including the deep seas. The pioneer submarine gravity measurements were made by Vening Meinesz (1929) using an ingenious three-pendulum system. Such submarine instruments, with an accuracy of 2–3 mgal, have been successfully employed for collecting gravity data at sea (Vening Meinesz, 1948; Worzel, 1965a). Subsequent developments have led to the design of various underwater and surface-water ship gravimeters (LaCoste, 1967), and in recent years satisfactory tests of an airborne gravimeter have been reported. Brief descriptions of such special gravity measuring instruments can be found in Nettleton (1971).

Field operations

Location and spacing of stations. One of the most important considerations in a gravity survey is the location and spacing of stations. Wherever possible the gravimeter stations are planned at the corners of a square grid, the grid length, s, depending primarily on the depth, h, of the geological feature to be located. An easily remembered rule is that s should be $\lesssim h$. In large-scale surveys, covering some thousands of square kilometres for the purpose of mapping regional geological trends and major crustal structures, measurements are carried out by vehicle or helicopter (or by a ship in marine areas) with a station spacing, s, ranging from a few kilometres to tens of kilometres. On the other hand, in small-scale surveys for detailing local structural features, s may be between some tens of metres to hundreds of metres. In a reconnaissance survey for oil exploration s is usually of the order of 1 km.

It is always advisable to avoid placing the stations near topographic features which might have a significant effect on the gravity readings. This should be done mainly to minimise the corrections that would be necessary to account for the gravity effect of a topographic mass.

Levelling survey. The geographical positions and the elevations of gravimeter stations must be accurately known in order to reduce the gravity readings to sea level or to any arbitrary datum level, as is described later. This requires a considerable amount of topographic surveying. Spirit levelling is usually necessary, but in some cases barometric levelling is adequate. Some geological reconnaissance must also be done, together with the collection of unweathered rock samples, so that data on rock density can be made available for the elevation and topographic corrections. Marine gravity surveys require elaborate radio or radar systems to determine station locations.

The field routine for gravity measurements is similar to that for an elevation survey. Gravity differences, Δg, are determined for a network of stations with respect to a starting point (base station) after allowing for drift (see p.108). Closure errors of the network are adjusted by least squares, in much the same way as is done with geodetic level nets. The end result of field operations is a network of points for which Δg has been determined. Before being compiled into a usable map, these values must be corrected for certain well-known effects.

Reduction of gravity data

The field data of gravity differences between an arbitrary reference point and a series of stations is subject to various extraneous effects which are unrelated to the subsurface geology. For instance, effects due to the oblateness of the earth, and to changes in elevation and topography must be carefully removed from the observed data before any geophysical interpretation can be seriously attempted. This process of correction, or so-called reduction of data, is now a well-established routine in any gravity survey. The necessary corrections are: (1) latitude correction; (2) free-air correction; (3) Bouguer correction; and, sometimes, (4) terrain correction.

Latitude correction. This correction is made to remove the effect of the increase of gravity from the equator to the poles as mentioned previously. The basis of the correction is the I.U.G.G. (1967) Gravity Formula (see p.93). Published tables of the normal sea-level gravity are available from which the corrections are readily determined from the latitudes of the stations as shown on the station location map. Usually the correction is made to an arbitrary base latitude ϕ. If a survey is of a rather limited extent covering north–south distances of only a few kilometres, it may be sufficient to use the following correction factor:

$$C_\phi = 0.812 \sin 2\phi \text{ mgal/km (north–south)} \qquad (3.22)$$

It must be subtracted from, or added to, the measured gravity difference, depending on whether the station is on a higher or lower latitude than the base station. For mid-latitudes the correction is about 0.08 mgal/100 m, which gives an idea of the accuracy needed in knowing the relative positions of stations.

Free-air correction. This correction takes care of the vertical decrease of gravity with the increase of elevation. The basis of the correction is the inverse square dependence on the distance in Newton's law (see eq.3.3). If g_0 be the gravity at the datum level (usually but not necessarily the sea level), then at a height, h, above it, the gravity would be:

$$g = g_0 \frac{R^2}{(R + h)^2} \simeq g_0 (1 - 2h/R) \qquad (3.23)$$

where R is the earth's radius.

For small elevation differences, the free-air correction is:

$$C_{\mathrm{F}} = 2g_0 h/R = 0.3086h \text{ mgal} \qquad (3.24)$$

where h is the elevation in metres.

The correction must be added to a measured gravity difference if the station lies above the datum plane and subtracted in the opposite case. For gravity differences to be accurate within 0.1 mgal, the elevation differences between stations must be known within one-third of a metre.

Bouguer correction. This correction takes into account the attraction of the material between a reference elevation and that of the individual station. This attraction can be approximated by treating the intervening rock material as an infinite horizontal slab, of a thickness equal to the elevation difference, *h*, between the reference base and the station. The gravity attraction of such a slab is traditionally known as the Bouguer effect (after Pierre Bouguer, who first discovered this effect during the course of his measurements in the Andes), and is given by:

$$C_B = 2\pi G\rho h = 0.0419h\rho \text{ mgal} \qquad (3.25)$$

where *h* is in metres, and ρ, the density of slab material, in g/cm³.

Referring to Fig.57, it is easy to see that the attraction due to the extra mass of the slab will make a positive contribution to the observed gravity at the higher station such as *P*. The Bouguer correction is therefore of opposite sign to that of the free-air correction. For a typical density of crustal material, 2.67 g/cm³, C_B is 0.1118 mgal/m. This is less than the free-air gradient (0.3086 mgal/m); consequently *g* measured on the land surface does decrease with increasing height at a net rate of about 0.2 mgal/m. The Bouguer correction is not

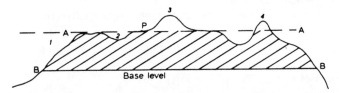

Fig.57.Bouguer and terrain corrections. The Bouguer correction refers to the gravity effect of the intervening mass between station *P* and base level *B*. The terrain correction takes into account the effects of topographic rises (3, 4) and depressions (1, 2). (After Parasnis, 1973.)

adequate for stations located in areas of rugged terrain and has to be supplemented by an additional correction for the departure of the terrain from a plane.

Terrain correction. This correction takes into account the attraction of topographic features like hills and valleys lying in the vicinity of a station such as *P* (Fig.57). Hills rising above the level of *P* will give an upward component of attraction, tending to reduce the gravity attraction caused by the rest of the earth. Any valleys below the station *P* correspond to holes (negative masses) in the Bouguer slab which also tend to reduce the gravity effect at *P*. The terrain correction C_T in both cases is positive, and should be added to the measured gravity difference. Calculation of the attraction of irregular topographic features is greatly facilitated by the use of special correction charts and tables (Hammer, 1939; Haalck, 1953). In recent years, various methods of calculating terrain corrections using fast computers have been devised (e.g. Kane, 1962).

When gravity is measured on the sea surface above the level of the sea bottom, the submarine terrain effect due to the bottom topography is quite different. The Bouguer and terrain effects are calculated as if water were replaced by rock. An example of the application of submarine terrain correction can be found in Nettleton (1971).

"Free-air" and "Bouguer" anomalies

In geodetic studies and regional surveys, it is customary to speak of gravity anomaly as the difference between the observed value of *g* at some point, and a theoretical value predicted by the I.U.G.G. (1967) Gravity Formula (eq. 3.7, p. 93), for the same point. The observed gravity value is determined by relative gravity measurements (made by gravimeters) with reference to certain primary base stations where absolute measurements have been previously made. If g_{obs} is the observed value on the land surface at height *h*, it must be corrected to sea-level before it can be compared to g_ϕ, the theoretical value for the same latitude.

If only the free-air correction, C_F, has been applied, we define the "free-air" anomaly, Δg_F, as:

$$\Delta g_F = g_{obs} + C_F - g_\phi \qquad (3.26)$$

Occasionally, the gravity data are presented as a "free-air" anomaly map without taking into consideration the topography above sea-level. This usually occurs when maps of oceans and continental shelves are drawn; any free-air map from land observations will show a strong correlation with local topography. In the past most gravity measurements at sea were made with the Vening Meinesz pendulums in submerged submarines. For these stations, the observed gravity must be corrected upward to sea level, by the subtraction of C_F and the addition of a term equal to twice the attraction of the water layer above the submarine, i.e. $4\pi G\rho d$, where ρ is the density of seawater, and d is the depth of the submarine.

In the case of gravity observations on land, it is usual to apply the free-air correction (C_F), the Bouguer correction (C_B), and the terrain correction (C_T) in order to reduce the observed gravity g_{obs} to the reference geoid (mean sea level). The Bouguer anomaly Δg_B is then expressed as:

$$\Delta g_B = g_{obs} + C_F - C_B + C_T - g_\phi \qquad (3.27)$$

This anomaly should be zero if there were no horizontal variations in the density of rocks below sea level. A Bouguer anomaly different from zero may represent the effect of lateral variations in density of rocks below sea level, or it may show that the actual density above sea level differs from that assumed in the calculation of Bouguer and terrain corrections.

In small-scale gravity studies, such as those related to local structural or prospecting problems, the reduction of data is usually made with reference to any convenient base point to which an arbitrary gravity value is given. To observed gravity differences, Δg_{obs}, from this point the necessary corrections (described in the previous section) are then applied. The resulting anomaly values will differ from the true Bouguer anomaly values only by a constant amount.

Field determination of density

Both the Bouguer correction and the terrain correction require a prior estimate of the density of the surface material within the range of elevation differences in the area surveyed. Direct sampling of surface rocks over a large area is a formidable task, and the density estimate so made may not even be representative of the rock material lying at moderate depths. Therefore, various field methods have been devised for the determination of density in situ. One of the common methods

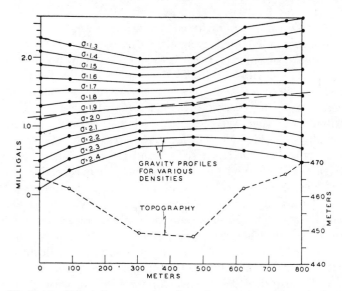

Fig.58. A method for measuring the surface density with a gravimeter. This profile indicates a density (σ) of 1·9 g/cm³. (After Nettleton, 1940).

is to run a "density profile" (Nettleton, 1939) with a gravimeter. This consists of measuring a line of closely spaced stations over a topographic feature. The observed gravity values are then reduced with different density values. The criterion for the actual density (Fig.58) is that which gives the smoothest reduced curve for the Bouguer anomaly and shows no visible correlation with the topography. The method has the advantage of sampling a comparatively large mass of rock material.

A tool often employed for direct measurement of formation densities in boreholes is the "density logger" which utilizes the principle of gamma-ray scattering (Baker, 1957). Recently density measurements at depths have also been made using a borehole gravimeter (Howell et al., 1966).

ISOSTASY AND CRUSTAL THICKNESS

In the early history of gravity measurements, it was found that the Bouguer anomalies, even after careful removal of the effects due to elevation and visible topography, show a systematic correlation with topography over broad regions. In high areas the Bouguer anomalies are almost always negative, whereas over the oceanic areas the anomalies are strongly positive. Over land, very near sea level, the average Bouguer anomaly is close to zero, but for broad areas with large relief these anomalies can reach some hundreds of milligals. The existence of such anomalies can only mean that beneath high areas the rocks are of below-average density and beneath ocean basins they are of above-average density. This is an exemplification of the important principle of the compensation of the earth's surface features by subsurface mass distributions. As this principle was suggested long before detailed measurements of gravity were available, it is interesting to study the historical development of the idea of "mass compensation".

The concept of isostasy

It is claimed (Delaney, 1940) that Leonardo da Vinci, the versatile genius and author of the 16th century, was the first to realize that the visible masses of the earth's surface are in equilibrium. Bouguer (1749) and R.J. Boscovich came much later to the same conclusion. However, definitive ideas about the mass compensation under the mountains, and explanations as to how such massive topographic features can be supported by the earth, came later in the 1850's from the analysis of deflections of the plumb line during a geodetic survey in North India.

If the Himalayas were really an extra load heaped on a rigid substratum, the plumb line (Fig.59) should be deflected towards the mountain range by an amount corresponding to the surface load represented by the mountain. However, Pratt's measurements (1855) showed that the observed deflection was much smaller, about one-third of that expected.

Two different hypotheses were advanced almost simultaneously to explain these observations in terms of the underlying mass deficiency. According to Pratt, the Himalayas

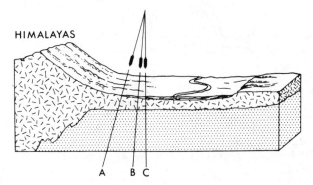

Fig.59. Deflection of the plumb line by the Himalayas: A = theoretical deflection that should be caused by the mountain mass, B = observed deflection, which is distinctly less because of the presence of the mountain "root", C = undeflected position.

have uplifted in such a way that the higher the mountain, the smaller its mean density. He generalized his theory by proposing a crust, extending to a uniform depth below sea-level, in which the density varied as the height of the topography (see Fig.60). As a geological justification for this, he postulated that mountains were formed by a vertical expansion of crustal columns, with no change in mass. Thus, all crustal columns have the same mass above a uniform level, the so-called "depth of compensation".

According to Airy's (1855) compensation theory, the Himalayan range was underlain by a "root" of light material so that the total mass beneath the mountain structure was no greater than that beneath the adjacent low-lying land. Also according to his theory, the higher the mountain, the deeper its root must penetrate into the heavier substratum. This would be possible if the substratum behaves like a "fluid" and the lighter mountain mass is floating on a heavier fluid substratum, somewhat like an iceberg floating on water. The depth of compensation is, thus, variable and the base of the crust in general follows an exaggerated mirror image of the real topography (Fig.60).

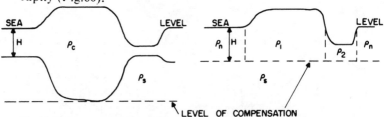

Fig.60. Isostasy: assumptions of Airy (left) and Pratt (right). In Pratt's (1855) model, the earth's crust extends to a uniform depth; the density varies with the height of the topography. In Airy's (1855) model, the base of the crust follows an exaggerated mirror image of the topography. (After Garland, 1971.)

Thus, the studies of the Himalayas formed the origin of the concept of isostatic compensation, although the term "isostasy" was first introduced in 1889 by the American geologist Dutton to describe the state of hydrostatic balance below a certain depth within the earth. In simple words, isostasy is merely the application of Archimedes' principle to the uppermost layers of the earth. As we shall see later, the existence of isostatic movements affecting the crust shows that a lateral flow must be able to occur in the relatively weak region below the depth of compensation, which is commonly called the "asthenosphere". In contrast, the overlying, relatively strong, "lithosphere" must reach equilibrium either by elastic bending or by a combination of fracture and flow.

Testing isostasy by gravity measurements

Some decades later a new phase in the study of isostasy began with the measurements of gravity over continents and oceans. The general trend of Bouguer anomalies, Δg_B, being largely negative over mountainous regions (for the Alps, $\Delta g_B \sim -110$ mgal) and strongly positive over oceanic areas (for the East Atlantic, $\Delta g_B \sim +270$ mgal) did not cause much surprise. The supporters of isostasy had already predicted that the mean density of the earth's curst is smaller under the mountains, and larger under the oceans, than under the flat low-lands.

Of particular importance is the study of the regional variations in gravity and their relationship to visible topography and crustal thickness. Fig.61 shows the relation of Bouguer anomalies to surface elevation for $3° \times 3°$ areas on the different continents (Wollard, 1970). The marked spread of about 80 mgal in Bouguer anomaly values for any elevation should have its origin in the mass distribution below sea level; this implies that there are significant regional changes in the mean crustal (and/or upper mantle) density. Gravity data alone cannot yield a unique solution for the subsurface density distribution. This is because an infinite number of mass distributions can create the same gravity effect (a typical example is shown in Fig.70). As seismic measurements give a detailed picture of the outer layers the earth, isostatic investigations should

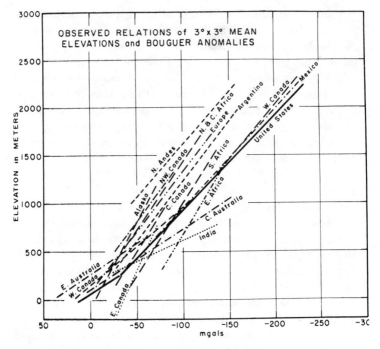

Fig.61.The relation of $3° \times 3°$ mean Bouguer anomaly values to the mean elevation on a worldwide basis. (After Wollard, 1969.)

be correlated with these wherever possible.

Depth of compensation and the "Moho"

The Mohorovičić discontinuity (in short the "Moho") is defined as the boundary between the crust and the mantle, and is characterized by a sharp increase in seismic velocity, V_p, from about 6·5 km/s above the boundary to a little over 8 km/s below the boundary. From the general relations of velocity and density (see Fig.9, p.20) and also from considerations of the variation in density with depth inside the earth, there are strong indications that this boundary is also one where a sharp density contrast occurs, with probable density values of roughly 2·9 above and 3·3 g/cm³ below the Moho. If Airy-type isostatic compensation is effective, the base of the seismological crust as depicted by the Moho should be an exaggerated mirror image of the surface topography. That this applies on a worldwide basis is confirmed by the seismological results shown in Fig.62. There is no doubt that the Moho discontinuity mirrors the topography regarding the differences between the oceans and continents. Within the continents, the greatest crustal thickness is observed beneath the Academy mountains of the U.S.S.R. The crust is thinnest beneath the deepest oceans, and tends to thicken beneath the mid-ocean ridges and oceanic islands. The magnitude of the Bouguer anomalies (in Fig.62) reflect the extent of the compensation in terms of the thickness of the low-density crust. The isostatic anomalies, however, remain much closer to zero over both oceans and continents, indicating that the crustal thickness variations are close to those required for isostatic compensation.

It is apparent from Fig.62 that there are some regions which are either under- or overcompensated. For example, the Trans-Carpathian is a region of low relief, but the crust is thicker than normal for a continental area and the isostatic anomaly is distinctly negative. Conversely, for its depth of water the crust beneath the Western Pacific is thinner than normal, and the isostatic anomaly is positive.

Further, it must be emphasized that the general relationship between the topography and crustal thickness is not confirmed when a number of crustal sections from one continent are studied. Pakiser (1963) concluded from his investigations that changes in the crustal thickness across the boundaries of the major geological provinces of the western part of the U.S.A. bear little direct relation to the change in topography unless the seismic velocity remains constant across the boundary.

No single universal hypothesis of isostasy can explain all the major surface features of the earth. A more complete explanation of the mechanism of compensation must await

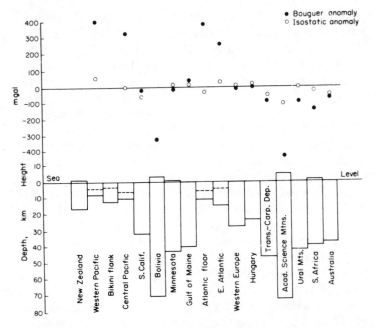

Fig.62.Crustal sections determined by refraction seismology over oceans and continents. Note the correlation between the topography, Bouguer anomaly and crustal thickness. (After Garland, 1971.)

The isostatic anomaly is computed from the Bouguer anomaly by applying an additional correction for the gravity affect of crustal blocks extending to various depths below sea level.

further investigations of the crust, both by seismic and gravity methods. In particular, the regional variations in upper mantle properties have to be studied. With the present information it can be said that lateral density variations, both above and below the seismological boundary (the Moho), appear to be required to supply the compensation in those areas where the Airy mechanism alone is not sufficient.

Nature's isostatic experiments

It is well known that the elevation of the land surface is rapidly changing over some areas of the earth. For example,

at the present time in Fennoscandia and northern Canada the
land surface is rising at a maximum rate of about 1 cm/year.

Let us take the interesting case of the land uplift in Fennoscandia. By measuring the elevation of ancient shore lines of
different parts of Fennoscandia, it has been possible to determine the amount, and also the rate, of land uplift. The general
opinion of the Quaternary geologists is that the land surface
in this area has risen about 500 m since the Pleistocene glaciation. The geodetic method for measuring the rate of uplift is
the most accurate. Fig.63 shows the rate of uplift obtained
through precise levellings made at intervals of about 30–40
years. The uplift in the middle part of the Gulf of Bothnia
amounts to about 100 cm per century. This value gradually

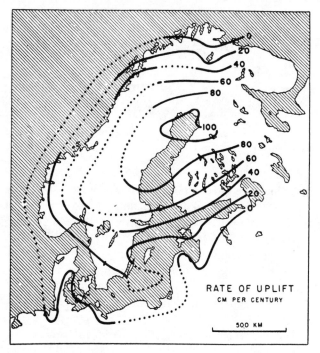

Fig.63.Postglacial uplift of Fennoscandia. (After Gutenberg, 1959.)

124

falls towards the periphery. Beyond the zero contour the land is sinking, although rather slowly.

Why this uplift and sinking of the land? The phenomenon can be explained as follows. During the last ice age, Fenno-scandia, loaded by thick ice caps ($\sim 2 \cdot 5$ km thick), had sunk as much as 600–700 m in the mantle. Since the ice melted away (about 10,000 years ago), Fennoscandia has been steadily rising in accordance with the principle of hydrostatic balance. The uplift of the central area and subsidence near the periphery still continues today. In other words, we have here "living" evidence of isostasy. An important outcome of this evidence is that at present the flow of the subcrustal material is directed toward the central uplift region. Fig.64 illustrates

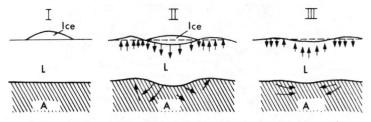

Fig.64. Schematic representation of the response of the asthenosphere (A) to the application and removal of a load, such as an ice cap, on the surface of the elastic lithosphere (L).

the point. Since the subcrustal mass that moved sidewards during the glacial period has not had time to move back (due to the high viscosity of the mantle material), there is still too little mass under the centre, and the gravity anomalies are, therefore, negative being as high as -50 mgal over the Gulf of Bothnia (Fig.65). From the amplitude of negative anomalies along several profiles, Niskanen (1939) estimated that the land had still about 200 m to rise before compensation is restored.

Other examples of nature's isostatic experiments can be seen in Greenland and in Antarctica. The surface of these land masses was obviously above the sea level before the large

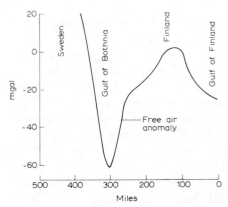

Fig.65.Gravity anomaly profile from Fennoscandia. (Based on Honka-salo, 1960, and redrawn from Garland, 1965.)

ice sheets covered them. Later, because of the heavy ice load, the land surface sank below sea level to preserve the isostatic equilibrium. Fig.66 shows the region of central depression in Greenland and the thickness of ice determined from seismic measurements. Thus, in Greenland and Antarctica we see the first, and in Fennoscandia, the second, phase of nature's experiment.

Isostatic rebound and viscosity of the mantle

The restoration of isostatic equilibrium in an area where

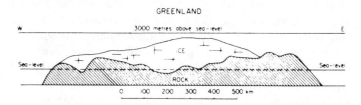

Fig.66.The ice sheet of Greenland. The profile shows the thickness of the ice sheet as determined by seismic measurements. Note the depression of the basement below sea level by as much as 250 m.

crustal movements have occurred is a matter of considerable interest in connection with the rheological properties of the upper mantle. The history of land uplift in Fennoscandia is so well known through precise levelling of the dated shore lines, that the variation in height with time can be used to determine the rate of recovery towards equilibrium.

The rate of recovery is conveniently measured in terms of the relaxation time, τ, which is the time required for the deviation from isostatic equilibrium to be reduced to $1/e$ (i.e. $1/2\cdot718$) of its initial value. It is necessary for these analyses to know not only the present rate of uplift, but also the amount of recovery still required. Fennoscandia has already risen about 500 m since the last glacial period and probably has a further 200 m to rise. The relaxation time is inversely proportional to the dimension of the load. It is about 3500–4000 years for Fennoscandia (Jensen, 1970, p. 179).

The rate of isostatic recovery, which is related to the rate of flow of mantle material, can be used to estimate the viscosity of the mantle. Using the observed rate of recovery, Niskanen (1939) estimated the viscosity of the asthenosphere below Fennoscandia to be $3\cdot6 \times 10^{22}$ poises (P). Heiskanen and Vening Meinesz (1958) estimated it to be 10^{22} P. For comparison, the viscosity of water is $0\cdot010$ P, that of thick honey is about 100 P, and that of flowing lava about 4×10^4 P.

The Hudson Bay area of northern Canada, heavily loaded with ice during the Pleistocene glaciation, is undergoing a similar postglacial uplift (Innes, 1960). The North Canadian and Fennoscandian ice sheets had similar areal dimensions and thicknesses. The North Canadian uplift also yields a viscosity estimate of about 10^{22} P. The close agreement with the value for Fennoscandia is to be expected because both regions are Precambrian shields with little evidence of any strong deformation since the beginning of the Cambrian.

It is, of course, very exciting to be able to calculate the viscosity of the upper mantle in this way, but these estimates should be interpreted with great caution. The estimates are based on the assumption of a Newtonian viscous fluid. It is possible that other types of creep behaviour in the upper mantle can give rise to the same observed phenomena.

Regions of uncompensated masses

Large isostatic anomalies, which indicate the presence of uncompensated masses on the earth, may throw light on the behaviour of the mantle under stresses of long duration or on the convective flow pattern proposed for the upper mantle.

The most striking example of large isostatic anomalies is that of the Archipelago of Indonesia. Narrow belts of intense negative anomalies (up to -200 mgal) are observed along the island arc. Fig.67 shows the isostatic anomalies associated with the ocean deep along the arc. In view of the large deviations from isostatic equilibrium in these belts and the fact that the islands situated in them show marked folding and overthrusting, Vening Meinesz (Heiskanen and Vening Meinesz, 1958) reasoned that the crust in these zones was subjected to strong lateral compression. He attributed the uncompensat-

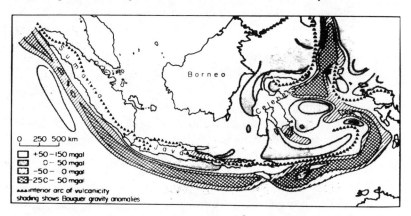

Fig.67.Isostatic gravity anomalies in Indonesia. The belts of negative anomalies were interpreted to be expressions of the downbuckling of light crust into the denser mantle. (Based on Heiskanen and Vening Meinesz, 1958; reproduced from Cook, 1973, by permission.)

ed mass deficiency to a downbuckling of the lighter crust into the denser mantle. The shape of the gravity anomaly, in particular the central location of narrow strips of negative anomalies, was interpreted to be suggestive of a symmetrical thickening of the crust. Vening Meinesz considered that the thickening resulted from a combination of downbuckling and plastic flow towards the buckled region by the drag of converging convection currents. Crustal downbuckles (also called tectogenes) suggested by the gravity anomalies seemed to form evidence of the importance of lateral compressional stresses in the earth. However, recent seismic investigations of deep-sea trenches have shown that in many cases crustal buckling alone is not an adequate explanation for the major part of the negative gravity effect (see p. 155).

The other striking example of an uncompensated mass feature is the island of Cyprus. Extremely large positive gravity anomalies have been known to exist over Cyprus. The gravity profile shown in Fig.68 represents the Bouguer anomaly, but over the island the isostatic anomalies are at least as large. The area is evidently one of great mass excess locally, the ultrabasic material representing an uncompensated excess load on the crust. Despite this excess load, the island has not sunk in historical time. The thick sections of Tertiary sediments now above sea level around its margins show that it has risen since Tertiary time. The geology of the island is rather complicated, as there are basic rocks of different ages. Most significant are the various exposures of olivine-bearing gabbros which are believed to be connected at depth. Relatively low-density sedimentary formations conceal the igneous rocks around the flanks of the island. The gravity data has been used by Gass and Masson-Smith (1963) to infer the distribution and depth extent of the ultrabasic rocks. Two of the plausible models that would explain the observed gravity are shown in Fig.68. For many years speculations have been prevalent that these ultrabasic masses are "windows" to the mantle, but there are uncertainties concerning the mechanism by which

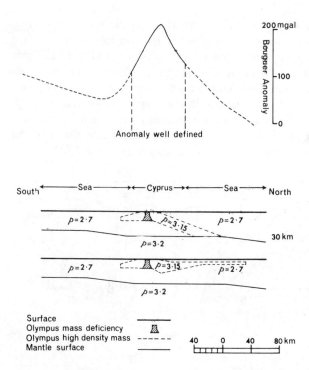

Fig.68. Gravity profile across the Troodos massif, Cyprus, and two models that could explain the observed gravity. (After Gass and Masson-Smith, 1963.)

the dense material was pushed upward through the crust. A recent suggestion based on seismic evidence is that this material is an upthrust slice of oceanic crust (Poster, 1973).

The presence of uncompensated masses may be taken as an evidence of long-term strength in the earth's mantle, if they are elastically supported, or of convection currents, if they result from the differences in density in a convecting fluid. In this connection it might be of interest to examine large-scale mass anomalies revealed by the satellite gravity observations. Major undulations of the geoid reflect widespread mass inhomogeneities within the earth (Fig.69). However, these large mass anomalies bear no obvious relationship to the surface mass distributions defined by continents and oceans. This is to be expected because major surface features are all in an approximate isostatic equilibrium. Thus, these anomalies cannot be caused by variations in the crustal thickness or density. It is also unlikely that a sufficiently large density contrast can exist in the lower mantle which would contribute to the geoidal anomalies of such large magnitudes (Fig.69).

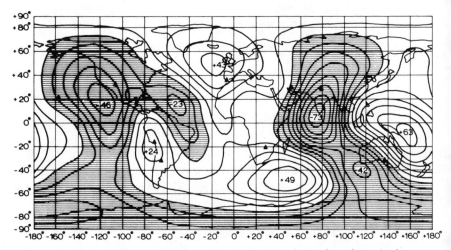

Fig.69.Sixth-degree curves of geoid heights at 10-m intervals as determined from satellite observations. A high geoid corresponds to relatively high-density rocks beneath, and a low geoid to low-density rocks (probably within the mantle). (After Iszac, 1965.)

The most likely explanation is that these anomalies are caused by lateral density variations within the upper mantle. A correlation, to some extent, can be seen between the geoidal highs and the areas of recent tectonic activity. For example, the geoidal high associated with the North Atlantic region centres over the Mid-Atlantic Ridge southwest of Iceland, which is believed to define a locus of crustal spreading. The geoidal high associated with the southwest Pacific centres over the New Guinea–Solomon Islands region, which appears to be a locus of crustal convergence. However, there are difficulties. If mantle convection is a factor, it is of opposite sign in the above-mentioned two areas. Further, no adequate explanation is yet available for the most prominent geoidal depression (gravity low) just south of Ceylon in the stable region of the Indian Ocean basin. Nevertheless, it seems possible that stress, tensional or compressional, in combination with temperature abnormality may result in a large-scale mass anomaly, the most likely place for it to occur is the upper mantle where anomalous mass can be created through a mineralogical phase transformation (see Fig. 191).

ANALYSIS AND INTERPRETATION OF GRAVITY DATA

The end product of a gravity field survey (after applying all the necessary corrections) is usually a contoured anomaly map, the so-called Bouguer map. The objective of gravity interpretation is to infer the geological character of the sub-surface structures from the anomaly map. To a geophysicist, positive and negative Bouguer anomalies mean much more than mere indications of mass excesses and deficits. The ultimate aim of the geophysicist is to deduce from the various characteristics (in particular the amplitude, shape, and sharpness) of the anomaly the location and form of the structure which produces the gravity disturbance. For this purpose, the data has to be analysed by suitable interpretation techniques. It is important for all who use gravity data—and in particular for geologists—to realize that despite the best use of the available techniques interpretation is not a clear-cut process which can be relied upon for a unique answer. The problem of magnetic interpretation (see Chapter 4, p.212) is essentially the same, and, therefore, the following discussion also applies to magnetic anomalies.

Ambiguity in gravity interpretation

There are two characteristics of the gravity field which make a unique interpretation almost impossible. The first is that the measured value of g, and, therefore, also the reduced anomaly, Δg_B, at any station represents the superimposed effect of many mass distributions from the grass roots down. A Bouguer gravity map is almost never a simple picture of a single isolated anomaly, but is practically always a combination of relatively "sharp" and "broad" anomalies whose "causative" sources are at different depths. Interpretation can, therefore, only proceed after the contributions of the different sources are isolated by various techniques which will be described later. No matter what technique is employed, the anomaly separation process suffers from the axiomatic fact that a "sum" cannot be uniquely divided into "parts" without the imposition of restricting conditions.

The second, and more serious, difficulty in gravity interpretation is that of determining the "source" from the "effect", which is the "inverse" problem of the potential field theory. For a given mass distribution (or simply a mass source), it is fairly easy to determine its gravity effect, but the inverse problem has no unique solution. For a given distribution of gravity anomaly on (or above) the earth's surface, an infinite number of mass distributions can be found which would produce the same anomaly. Fig.70 provides an example where a given gravity anomaly could be explained by any of the alternative mass distributions showing a fixed density contrast, $\Delta\rho$, with respect to the surrounding material. Fig. 71 illustrates yet another type of ambiguity arising due to the lack of information about the densities. If we assume that the anomaly results from a buried flexure or a geological fault, two interpretations of the "step" structure are possible and the ambiguity cannot be resolved unless the densities are reliably known.

The above considerations, at first sight, may present a rather pessimistic picture of the interpretation problem. However,

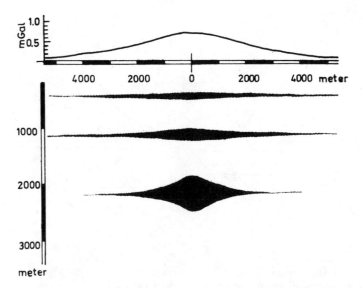

Fig.70. Ambiguity in gravity interpretation. The given gravity anomaly can be explained by any of the three lens-shaped bodies showing a fixed density contrast with respect to the surrounding material. (Modified from Griffiths and King, 1965.)

in actual practice the situation is not so hopeless. A geological intuition coupled with logic will often rule out solutions of many forms, and other information, such as the probable density or depth of the geological feature, may further narrow down the range of likely possibilities. Furthermore, if any independent control such as that obtained from drilling logs or seismic data is available, the number of variables can be reduced to the point where the final solution may have practical validity. Also, as we shall see later, some useful information on the estimates of maximum possible depth and total mass contrast can often be obtained from a gravity anomaly without making any assumptions about the causative body.

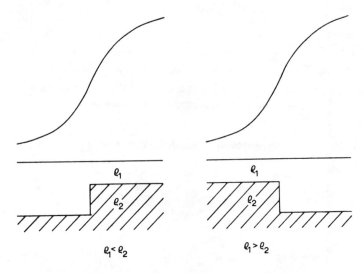

Fig.71.Alternative inferences from the same gravity profile in the absence of a knowledge of the density contrast. (After Parasnis, 1973.)

Isolation of anomalies

The gravity anomalies of relatively small geological features appear as minor distortions on the anomaly field due to broad crustal structures. These latter effects are perceptible over large distances, and are, therefore, called "regional" anomalies in contrast to the anomalies of small-scale structures which are termed "local" anomalies. The isolation of local anomalies is of prime interest in prospecting applications, whereas the regional anomalies are mostly used for crustal studies. There are several sources of regional gravity variation. In addition to large-scale geologic structures, there may be density effects caused by intra-basement lithologic changes. In certain areas the regional variation might be due to isostatic effects associated with deep-seated density anomalies. Difficulty often arises when several density anomalies, of either local or regional extent, occur so close together that their individual effects cannot be readily resolved.

In a gravity map of a small area, the regional trend may appear as a uniform variation represented by parallel, evenly-spaced contours. A local anomaly, which ordinarily would be indicated by closed contours, appears as a "nose" on the regional anomaly field (Fig.72). The anomaly separation procedure may consist of the removal of the regional effect by either of two methods: (1) graphical smoothing, either on the countour map or on profiles; (2) an analytical process applied to an array of values, usually on a regular grid.

In the graphical approach (which is more popular among geologists), the value of the visually smoothed field can be subtracted from the original Bouguer anomaly at the point to give the "residual" anomaly (Fig.72). While the process is

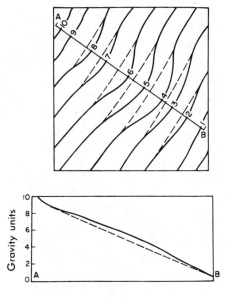

Fig.72.Graphical method of the separation of regional and local anomalies. The regional effect has been sketched by free hand (dashed lines) and on subtracting it from the observed anomaly, either on the contour map (top) or on the profile, the local anomaly (the so-called "residual") can be obtained. (After Garland, 1965.)

entirely empirical and subjective, it can, if skillfully applied, lead to a very effective isolation of anomalies.

The analytical approach is based on suitably averaging the gravity anomalies along a circle of specified radius around each station. The circle may be thought of as a band-pass filter, which is reasonably effective at isolating anomalies whose width is of the same order as its diameter. The average so obtained is the regional effect and subtracting this from the Bouguer value of the station gives the residual. The analytical technique, thus, appears to be free from subjective bias, but the process is too mechanical and known geological factors, which might affect the interpretation, may not be taken into consideration. The analytical methods can be readily programmed on digital computers, and are, therefore, much more rapid than the smoothing techniques. Further details on various techniques of anomaly separation can be found in a number of articles (e.g. Nettleton, 1954; Roy, 1961).

Analytic continuation of fields

By a mathematical process involving surface integration (see Parasnis, 1972, p.59) it is possible to calculate the form which a given potential field (e.g. a gravity or magnetic anomaly field) would have if it were measured at a higher or at a lower level.

Upward continuation always gives a smoother anomaly map than the original. It can be used as an effective smoothing technique for the separation of anomalies due to deeper sources. In airborne magnetic surveys, upward continuation has been used to calculate the field at a higher level and the computed results agree fairly well with measurements actually made at that level.

The function of downward continuation is to convert the observed field data to what it would be if the measurements were made to a level closely above the anomaly sources. Thus, it has a particular application in the resolution of overlapping effects of sources close together. However, if continuation is carried to depths greater than the depth of the major feature to be located, the continued field will begin to show extreme fluctuations. The level at which this fluctuation starts may, in certain cases, give a direct indication of the depth to the top of the anomalous feature producing the given field, but actually it provides a limit for the maximum acceptable depth. For further details of the continuation process and its applications, the interested reader is referred to papers by Peters (1949), Hammer (1963) and Roy (1966).

Estimates of depth and mass

In general the sharpness of an anomaly is an index for the depth of the causative source. If some regular shape for a geological body can be postulated (e.g. sphere, horizontal cylinder, semi-infinite slab), simple depth rules in terms of half-width or some other measure of the anomaly gradient can provide unambiguous estimates of depth. Some of these were already mentioned in the section *Gravity effects of simple mass shapes* (p.98). However, when no assumptions as to the shape of the anomalous body can be made a unique determination of depth is not possible.

Smith (1959, 1960) has derived some maximum-depth rules regardless of the shape of the anomalous mass distribution. Two of these are briefly discussed below:

(1) If Δg_{max} and $\Delta g'_{max}$ are respectively the maximum values of the gravity anomaly and its horizontal gradient, then the depth Z_T to the top of the body must be:

$$Z_T \leqslant 0 \cdot 86 \mid \Delta g_{max}/\Delta g'_{max} \mid \qquad (3.28)$$

(2) When the anomaly is only partly known, a depth estimate is still possible by using the gravity value $\Delta g(x)$ and its horizontal gradient $\Delta g'(x)$ at the same point. In that case, the depth relation is :

$$Z_T \leqslant 1 \cdot 5 \mid \Delta g(x)/\Delta g'(x) \mid \qquad (3.29)$$

For two-dimensional mass distributions (bodies elongated in one direction) the numerical factor may be replaced by 0·65 in eq. 3.28 and by 1·0 in eq. 3.29.

These rules apply only if the involved density contrast has the same sign for all the sources contributing to the anomaly. They are especially useful for classifying anomalies thought to be due to flat-lying structures. A more complete discussion of the Smith rules can be found in Parasnis (1972).

The magnitude of a gravity anomaly is a direct measure of mass, and the total anomalous mass can be uniquely determined without any assumption whatsoever about the shape, density, and depth of the anomalous body. The basis of calculation is a Gaussian surface integration of the residual anomaly (i.e. Bouguer — regional) over the area of measurement. The formula for the total anomalous mass (see Parasnis, 1972, p. 63) is:

$$M = \frac{1}{2\pi G} \iint_{S} \Delta g \; \Delta S = 23 \cdot 9 \; \Sigma \; (\Delta g \times \Delta S)$$
$$\text{metric tons} \quad (3.30)$$

where Δg is the mean anomaly (mgal) within a small area element $\Delta S \, (\text{m}^2)$ and Σ denotes the summation of the products $\Delta g \times \Delta S$ over the whole area of measurement.

The above relationship is of great practical importance in mining geophysics. The calculation of the actual ore mass, however, requires that the ore density, ρ_A, and the density of surrounding rock, ρ_B, be known. The actual ore mass is then simply $M \times \rho_A/(\rho_A - \rho_B)$.

Interpretation by models

Almost all the interpretation of gravity data is done by indirect methods, as there is no direct method of translating gravity data into subsurface geology. The general procedure is to assume the various simple shapes (in accordance with geological plausibility) for the source of an anomaly, compute their gravity effects at the surface, and modify these models progressively until a reasonable fit with the observed anomaly is obtained. The closeness of the fit indicates only that the selected model is a possible solution. If there is insufficient geological data it may not be possible to make more than "trial and error" attempts to select a range of approximate solutions from which the most likely solution is to be sought.

Before choosing a model for the trial calculations, a careful inspection of the isogal anomaly gap is necessary to extract some clues about the size and form of the causative geological body. In particular the elongation of the isoanomaly contours is a useful index of the direction and length of the structure causing them. When choosing a model in the light of geological information available for the area, it is usual to make the simplest geometrical approximation for the model shape. The use of complicated models only to obtain a better fit between the calculated and observed anomalies may be pointless especially when there is no adequate geological or other independent control. Interpretation by use of simple models can best be illustrated by a field example. Fig. 73 shows the Bouguer anomaly observed over a local feature in the salt dome province of North Jutland in the Danish sedimentary basin. Salt bodies are usually characterized by gravity lows because salt ($\sim 2 \cdot 2$ g/cm^3) is normally less dense than the surrounding sediments. This anomaly is almost circular in form and the effect of the salt mass is so prominent that the regional effect is hardly perceptible. The total amplitude of the anomaly, Δg_{max}, is about -16 mgal, and the half-width, $x_{1/2}$, as determined from NE–SW profile is approximately 3700 m. Assuming a spherical form for the anomalous body the depth, Z_c, to the centre of the salt mass is given by the half-width rule* as $1 \cdot 305 x_{1/2}$ which is about 4800 m. From the total amplitude of the anomaly we can estimate the radius,

---

*Various half-width rules for estimating depth have been discussed in the section *Gravity effects of simple mass shapes* (p.98).

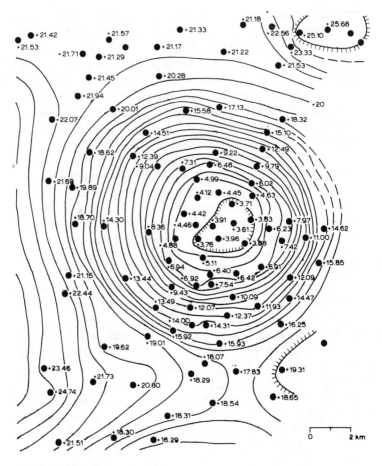

Fig.73.Part of the Bouguer anomaly map in the salt dome province of northern Denmark. (After Saxov, 1956.)

R, of the equivalent sphere. However, this requires some information about the average density contrast of salt with respect to the surrounding Cretaceous and older sediments at depths of the order of Z_c. If we assume an average density contrast of -0.25 g/cm³, then by using eq. 3.11, we obtain the value for R of 3800 m. The depth, z_T, to the top of the salt dome, given by Z_c-R, is then estimated to be approximately 1000 m.

Even without more specific information on the densities and on the depth of the dome at some point (such as that obtainable by a drilling contact or seismic data) the simple calculations with the spherical model have been able to provide geologically significant information about the location, the approximate depth, and a rough measure of the diameter which outlines the deeper flanks of the dome. The point of this example is to illustrate that there is no geological usefulness in modifying the simple spherical model to other shapes only to obtain a better fit between the calculated and observed anomaly curve, unless this is warranted by some additional specific information about the nature and depth of the density contrasts associated with the salt structure. The other point to be emphasized here is that the details of the gravity data which can be useful depends very much on the objective of the survey. The Bouguer contour map of Fig.73 is based on stations approximately 1–2 km apart and there are some wide gaps especially along the eastern margin of the area. The data is quite adequate for locating and determining the general outline of the salt dome. However, to determine the details of a dome with an associated shallow cap rock would require a much closer station spacing over the central part of the dome in order to map the relatively sharp changes there and to separate the cap-rock effect (usually showing a small positive anomaly) from the salt effect. The criterion for station spacing in relation to the depth of the feature to be mapped has already been mentioned earlier (p.109). An example of detailed gravity studies on a salt dome with a remarkably strong cap-rock effect (of over 4·5 mgal) is reported in Ramberg and Lind (1968).

Use of computational aids in interpretation

The elementary calculations based on approximating geologic structures to simple mass shapes (e.g. sphere, cylinder, slab) serve fairly well as an adequate control on geological assumptions and as quantitative checks on various geological possibilities to account for the observed gravity anomaly. However, in certain problems of interpretation detailed calculations are useful especially when a geological situation is partly well defined by drilling or known stratigraphy or by some other control such as provided by seismic studies.

There are a number of graphical devices and machine methods which enable more detailed calculations to be made rapidly. In general they are classified as two-dimensional and three-dimensional methods. Two-dimensional calculations are applicable to situations where the anomaly contours are elongated. Fig.74 illustrates a simple graticule that can be used for this purpose. Some graticules have also been devised for calculating

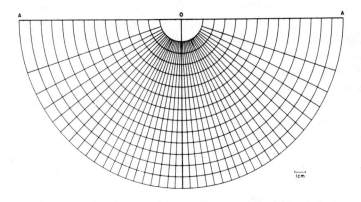

Fig.74. A graticule for the calculation of the gravity anomalies of two-dimensional bodies. The effect of each compartment is $6\cdot67 \times 10^{-6}\ \rho S$ mgal, where the scale is $1/S$, and ρ is the density contrast in g/cm³. (Modified from Griffiths and King, 1965.)

the gravity effects of three-dimensional bodies of arbitrary shape. Alternatively, a nomogram, based on the principle of the summation of the effects of rectangular laminae, can be used (Sharma, 1967). Nowadays, however, detailed calculations are usually made on digital computers. Of the various computerized methods, the one based on the approximation of the body by a number of thin polygons has been most commonly used. Details of this method can be found in the paper by Talwani and Ewing (1960).

We shall illustrate the use of detailed gravity calculations by a field example. The example chosen is that of a gravity survey of the Blue Glacier situated on the north side of Mt. Olympus in Washington. This glacier has been the site of much glaciological research on the attitude and mechanism of glacial flow for which the knowledge of bedrock configuration is very important. Because of the considerable local relief of the glacier bed the seismic survey did not provide good results, although usable reflections were obtained in some parts of the glacier. Therefore, a gravity survey was conducted to determine more precisely the depth and shape of the glacier.

Fig.75 shows the Bouguer anomaly map of the Blue Glacier. The density of the glacier ice was taken to be 0.90 g/cm^3. The mean density of 33 bedrock samples (representing low-grade metamorphic rocks of Mesozoic and early Tertiary age) was determined to be 2.73 g/cm^3. From the known upper shape of the glacier it was evident that a two-dimensional model could not be used. A three-dimensional model, based on the summation of the effects of a number of polygonal plates of varying dimensions (after the method of Talwani and Ewing, 1960) was used to simulate the gravity anomaly of the glacier and with the aid of a computer the best-fit configuration for the glacial bed was deduced. This is shown in Fig. 76 along with the seismic control points. The subsequent bore-hole tests confirmed the accuracy of the gravimetrically determined depths within 5–10%. Thus, the gravity data and the detailed calculations proved to be useful in providing an almost unique

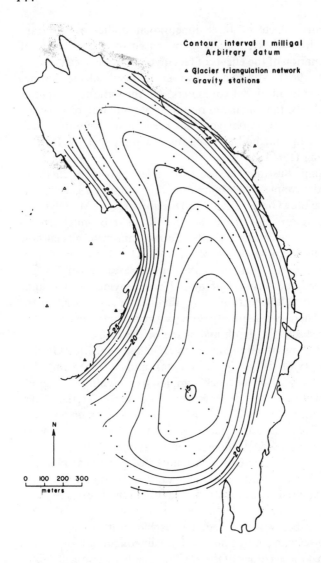

Fig.75.Bouguer anomaly map of the lower Blue Glacier, Mt. Olympus, Washington. (After Corbato, 1965.)

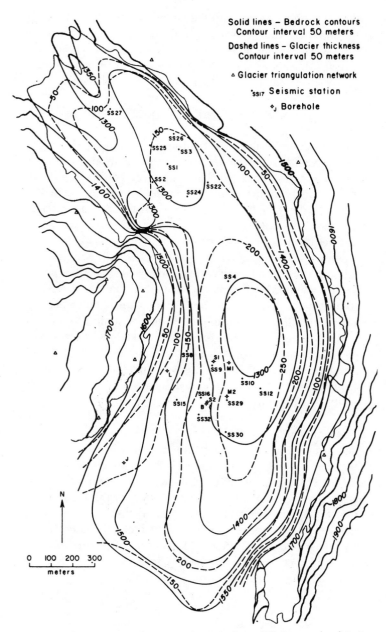

Fig.76. Bedrock contours based on the procedure of fitting gravity data to a three-dimensional model of the Blue Glacier. (After Corbato, 1965.)

solution mainly because of the simplicity of the shape of a valley glacier, the known density contrast, and the known upper surface. The point of this example is to illustrate that detailed calculations can be most useful for interpretation purposes when a geological situation is party well defined.

GRAVITY ANOMALIES AND GEOLOGICAL STRUCTURES

Gravity measurements have been used to study a great many types of geological structure, ranging in depth and size from very deep crustal blocks to near-surface ore bodies. In general, large regional variations in the Bouguer gravity are related to changes in the thickness of the earth's crust or are due to large-scale mass inhomogeneities. Local anomalous gravity values, on the other hand, are attributed to the near-surface mass inhomogeneities. Negative anomalies are identified with sedimentary basins, salt and granite structures, and grabens; positive anomalies are identified with uplifts, horsts, and mafic rock masses. It is possible to discuss here only a few examples which demonstrate the particular applicability of gravity studies to various problems related to regional and local structures. At the same time, the examples will indicate that, despite the problem of ambiguity in interpretation, gravity anomalies can give highly meaningful information about the subsurface structures and density distributions.

Granite and salt structures

The granite structures share with salt structures their peculiar character of being gravity minima. The negative gravity anomalies which have been observed over granites in many areas of the world (Cook and Murphy, 1952; Bean, 1953; Bott, 1956; Smithson, 1963) are significant in that they show the mean density of upper crustal basement rocks to be considerably higher (close to $2 \cdot 76$ g/cm^3) than that of granite (usually near $2 \cdot 67$ g/cm^3). When these anomolies were first noticed (Reich, 1932) they caused surprise, as it was thought that the upper part of the continental crust was typically granitic. From gravity evidence it is now very clear that the term "granitic layer" is a misnomer for the basement rocks of the upper crust (Bott, 1961).

In the case of granite bodies exposed at the surface, the gravity anomaly can be used to estimate the depth extent of the body and the attitude of its sloping contacts. Smithson (1963) has made detailed studies of some Precambrian granites in South Norway of which the Grimstad granite appears to be ideal for gravity interpretation. This is because the terrain effects are negligible, and both the outcrop pattern and the gravity anomaly field are simple. From geological considerations the Grimstad granite can be regarded as an almost circular (about 8 km in diameter) homogeneous granite that has near-vertical margins which are marked by an eruptive breccia and migmatic zone without thermal metamorphism.

Fig.77 shows the residual anomaly over a profile across the centre of the granite. The measured rock densities from surface outcrops reveal that a large density difference is found between the gneiss ($\rho = 2\cdot81$) and the granite ($\rho = 2\cdot64$), thus giving a density contrast of $-0\cdot17$ g/cm^3. Smithson made use of this density contrast to calculate a model that will simulate the effect of granite. A model of a vertical cylinder suggests itself because of the contact attitudes and the plan view of the granite. The solid-angle method (see eq. 3.16) was used to compute the gravity effect of models. The simplest model would be a single vertical cylinder with the same diameter as the granite; however, the asymmetrical shape of the anomaly rules out this possibility. Accordingly, two superposed cylinders of unlike diameter were employed as shown in Fig.77. The agreement between the observed and calculated anomaly curves is fairly good, and it suggests the minimum thickness of the granite to be about $2\cdot6$ km. The term "thickness" applied to a gravity model means the vertical extent of the density contrast. If the density contrast is assumed to decrease with depth the thickness of the granite would be considerably greater.

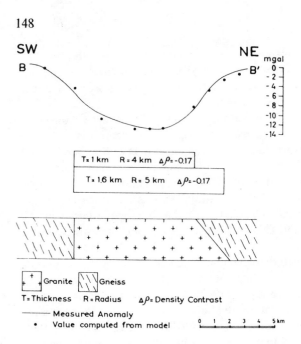

Fig.77.The residual gravity anomaly over the Grimstad granite, and the model calculated to simulate the gravity effect. (After Smithson, 1963.)

The other conclusion derived from the Grimstad anomaly is that the northeast contact of the granite dips under the gneiss much less steeply than is indicated by geological observations. A useful test to resolve the question of outward or inward slopes of contact can be applied as follows. Let A be the total amplitude of the anomaly over the granite intrusion. If the contacts are nearly vertical, the anomaly value over the contact will be about $A/2$. If the anomaly over the contact has fallen to less than $A/2$, then the contact slopes inward; if it is greater than $A/2$, it slopes outward. This test is very simple to apply and is generally valid provided that the body is broad in comparison with its depth and that the density contrast is uniform. Granite contacts characteristically slope outwards, and gravity studies of granites over various regions demonstrate that some really large plutons are thick (Dartmoor granite, England), that some large plutons are thin (Flå granite, Norway), and that some plutons are associated with large volumes of mafic rock (Skye granites, Scotland).

The gravity anomalies have been used not only for interpretation of the subsurface shapes of granites, but also for providing evidence for the problems of emplacement and origin of granites. A further consequence of the large negative anomalies over granite batholiths is that the granite could not have been produced by in-situ differentiation, because the settling of denser minerals would leave a layer of mass excess whose effect, if it were present, would have largely obscured the negative anomaly. In this respect, the other theory, based on the intrusion of a source magma of granitic composition, is more satisfactory, although the emplacement mechanism cannot yet be adequately explained. The granite case, thus, provides an excellent example of the direct application of gravity studies to geological problems. In an excellent review Bott and Smithson (1967) have discussed the granite problem on a geophysical basis. Recently Sorgenfrei (1971) has reviewed the main petrogenetic aspects of granite and has made an extensive compilation of data on granite and salt structures from several regions in order to show that granite and salt diapirism are closely related both structurally and genetically. Fig. 78 shows the gravity lows associated with the known salt basins in the North Sea area.

Rift valleys and sedimentary basins

Gravity surveys show that relatively large negative anomalies occur locally over the rift valleys. For instance, in the East African plateau some individual rift valleys show negative anomalies reaching down to about −150 mgal. Fig.79 shows the observed Bouguer anomaly profile over the Lake Albert rift valley (Girdler, 1964), and the computed model. The main problem of origin has been to determine whether the faults bounding the rift valley are normal or reverse. It has now

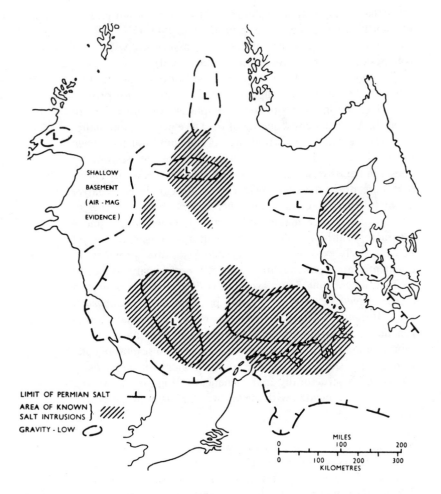

SHALLOW

BASEMENT

(AIR - MAG

EVIDENCE)

L

L

L

L

L

LIMIT OF PERMIAN SALT

AREA OF KNOWN ⎫
SALT INTRUSIONS ⎬

GRAVITY - LOW

MILES
0   100   200
0   100   200   300
KILOMETRES

Fig.78.Salt basins in the North Sea showing their relation to gravity lows.
(After Kent, 1968.)

been confirmed by both geological observations and gravity
profiles across the valleys that the faults are normal. This shows
that rift systems are caused by horizontal tension applied to

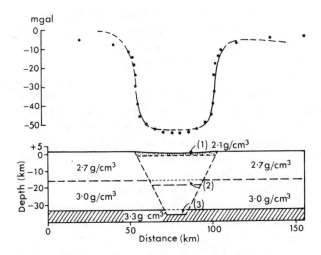

Fig.79.Observed Bouguer anomaly profile across the Albert rift valley. It agrees well with the computed model (solid dots) for a sediment-filled rift valley formed by normal faulting. (After Girdler, 1964.)

the crust. The older compression hypothesis has now been abandoned.

Fig.80 shows the Bouguer anomaly profile over the central part of the Rhine Graben rift system. A distinct asymmetry is

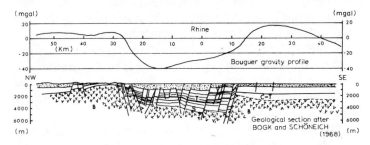

Fig.80.Bouguer anomaly profile across the central part of the Rhine Graben (Mueller, 1970). The geological section is from Boigk and Schöneich (1970): T=Tertiary and younger sediments, S=Bundsandstein, C T=pre-Permian sediments, B=basement rock.

evident from the anomaly observed within the graben, the axis of the gravity minimum being rather close to the western flank of the rift. Most of the negative anomaly can be accounted for by the low-density sedimentary fill in the graben, although the peak minimum observed close to the western flank does not fit well with the sedimentary cover shown in the geologic section. The negative anomalies cannot, therefore, be solely explained by the mass deficiency of the sediments in the Rhine Graben. Their cause must instead be sought at some depth within the crystalline basement. On the basis of the observed seismic velocity decrease and the associated density contrast, Mueller et al. (1969) suggested that at the corresponding depths a lateral change in the composition of basement material must take place. It appears that the general character of the graben basement is granitic ($\rho = 2.67$ g/cm^3) in comparison with the higher density ($\rho = 2.8$ g/cm^3) of the surrounding basement rocks.

Structure of coastal margins, oceanic ridges and trenches

Marine gravity surveys have proved a valuable tool in studying the deep structures of coastal margins, mid-oceanic ridges, and trenches. We shall discuss some examples of these studies.

Talwani et al. (1957) deduced a crustal section for the edge of the Bahamas in the vicinity of Eleuthra Island on the basis of gravity data (Fig.81). Here the continental margin crust abruptly thins to a normal oceanic crust within a distance of about 150 km; the major change occurs in the region of the 1000-fathom line. Note the sharp fall in gravity (> 120 mgal) within a distance of 100 km. The topographical slope is one of the steepest discovered in oceanic areas, and it provides one of the most striking examples of a continental section abruptly changing to an oceanic section.

The relation of gravity anomalies to crustal structure over the Mid-Atlantic Ridge has been discussed by Talwani et al. (1965). The minimum in the Bouguer anomaly (see Fig.82) over the crest of the ridge suggests an isostatic compensation of the ridge, but this can be either through a crustal root, or an

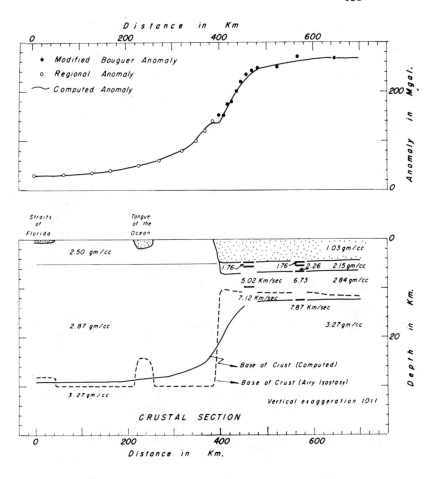

Fig.81.Crustal section across the continental margin at the Bahamas as deduced from the gravity profile. (After Talwani et al., 1957; reproduced from Worzel, 1965b.)

anti-root. The central depression in the Bouguer anomaly is certainly due to a low-density material extending to depths of at least 20–30 km as suggested by the computed models. The seismic evidence conclusively rules out the possibility of crustal thickening (crustal root). A possible density model that explains the gravity low over the ridge is shown in Fig.82.

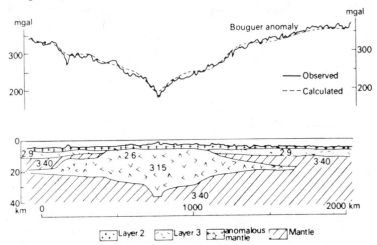

Fig.82.Gravity anomaly across the Mid-Atlantic Ridge. A possible density model that can explain the observed gravity low was deduced from the knowledge of seismic velocities. (After Talwani et al., 1965; redrawn from Bott, 1971.)

Gravity studies over several deep oceanic trenches have been reported by Worzel (1965b). Fig.83 shows the anomaly observed across the Aleutian trench which shows a marked drop in the free-air gravity of over 200 mgal. The geological model defining the crustal boundaries, as shown in the figure, was used for simulating the gravity effect. The overall agreement of the negative anomaly region is satisfactory. The positive anomaly in the vicinity of 280 km appears to be caused by mantle material of higher density extending towards the subcrustal region of the computed structure. The other positive anomaly (starting at about 450 km) with a wide spread demands a thinning of the crust. A similar thinning of the crust on the oceanward side of the trench is indicated by gravity and seismic measurements across other trenches, including the Puerto Rico, Peru–

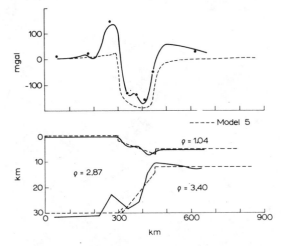

Fig.83.Gravity anomaly across the Aleutian trench and the proposed crustal model. (After Worzel, 1965b.)

Chile and Tonga trenches. These studies confirm that there is no need or indication of a tectogene (a large downbuckling of the crust deep into the mantle, see p.128) to explain the structural origin of trenches. The modern view, based primarily on seismological evidence, is that trenches are formed at places where a dense plate of lithosphere descends several hunderds of kilometres into the mantle. This idea is closely related to the tectonics of the ocean floor. It is further elaborated in Chapter 8.

Regional geology and tectonics

Regional gravity measurements yield information on the major structural elements and provide an excellent background for studying the broad architecture of a geological province. As an example, a simplified gravity map of the North Sea area is presented in Fig.84. On the basis of regional gravity minima, three sedimentary basins are outlined in the North Sea. These are referred to as the Northwest German basin.

156

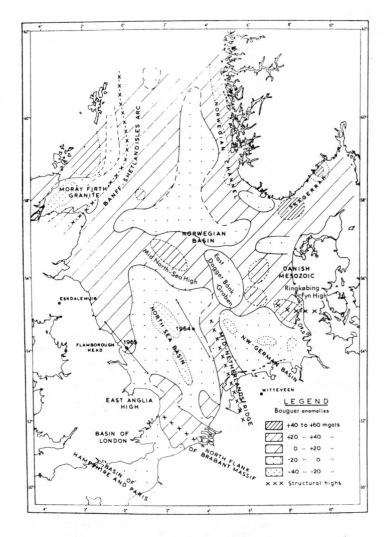

Fig.84.Simplified gravity map of the North Sea. (Adopted from Collette, 1968, with some additional data from Sorgenfrei, 1969.)

the British North Sea basin, and the Norwegian–Danish basin. The three structural highs, the Ringkøbing–Fyn high, the mid-North Sea high and the mid-Netherlands ridge, serve as dividing lines separating the three basins. The influence of salt structures located in the sub-basins is visible in the gravity map shown in Fig.84 (see also Fig.78).

The mean Bouguer gravity over the British North Sea basin and the Northwest German basin is of the order of −25 mgal. This should be seen against positive anomalies of the order of 20–30 mgal over the highs mentioned previously. The difference of 45–55 mgal can be accounted for by mass deficiency of sediments of probably 4500–5500 m thickness. This estimate follows from the gravity formula ($2\pi G t\ \Delta\rho$, see p.101) for an infinite slab of thickness t, assuming a density contrast of about $0\cdot25$ g/cm^3 between the basin sediments and the bordering basement rocks.

In a detailed map compiled by Sorgenfrei (1969), the two gravity highs are separated by a north–south trending graben which is called the East Dogger Bank graben. The other significant structural features are the London–Brabant massif with a positive anomaly, the Moray Firth gravity low and a local gravity high in the Skagerrak. As a crustal unit the London–Brabant massif may be comparable to the Ringkøbing–Fyn high, both being indicative of a shallow Precambrian basement. The Moray Firth low has been interpreted by Collette (1968) to be due to a granite batholith. In the northern part of the Skagerrak the continuation of the Oslo graben trend of gravity high is easily recognizable. On the other hand, the local gravity high in the Skagerrak in conjunction with the magnetic evidence (Sharma, 1970) has been interpreted to be indicative of a buried volcanic mount of probably early Tertiary age. The regional gravity low in the Skagerrak suggests the continuation of the sedimentary cover of North Denmark almost to the shores of Norway.

CHAPTER 4
Magnetic Methods

INTRODUCTION

Magnetic methods have a long history behind them. Early studies of the magnetism of rocks started with the discovery of the "lodestone" (magnetite-rich rock). It is generally believed that the Chinese were the first to make use of the directional properties of lodestone, almost certainly several centuries B.C. However, the idea that the earth itself acts as a magnet was realized much later, and in Europe the first reference to the use of a magnetic compass for navigation was toward the end of the twelfth century.

In 1600, with the publication of William Gilbert's book *De Magnete*, the concept of the earth's magnetic field and its directional behaviour was put on a scientific footing. Systematic quantitative observations of the direction of the earth's magnetic field were begun in London during Gilbert's life-time. It is claimed (Carlsborg, 1963) that observations of the local anomalies in the direction of the earth's field were in use in Sweden for iron-ore prospecting, as far back as 1640. However, it was not until 1870 that a special instrument was developed by Thalen and Tiberg for routine use in prospecting surveys.

Later with the growing use of the magnetic method of prospecting, the study of the magnetic properties of rocks became increasingly important. In particular, the discovery of the specific property of "remanent magnetization" in rocks and its subsequent applications to many exciting problems marked a turning point in the development of rock magnetism (including palaeomagnetism) so much so that since the 1950's it has become one of the most widely studied subjects in geophysics. In a short period of only two decades, palaeomagnetism has caused a revolution in geophysical thinking by providing quantitative evidence about phenomena of global significance such as continental drift and sea-floor spreading. It is doubtful whether the currently popular concepts of sea-floor spreading and plate tectonics would be taken seriously if palaeomagnetism had not already established the respectability of the continental drift as a scientific hypothesis.

The scope of the magnetic methods, encompassing the fields of geomagnetism, rock magnetism, palaeomagnetism and magnetic prospecting, is at present so wide that it is virtually impossible to cover all the aspects of their applications in a book of this size.

In this chapter, we shall briefly study the fundamental concepts in rock magnetism and geomagnetism, and discuss some of the important aspects of magnetic surveying and palaeomagnetic studies in geological investigations.

PHYSICS OF MAGNETISM AND MAGNETIC MINERALS

In studying the magnetic behaviour of bodies, including that of rocks, the various magnetic properties can be described in terms of certain fundamental quantities which have definite physical and mathematical meanings. In this section an attempt has been made to explain the fundamental concepts of magnetism with an emphasis on the physics of magnetic minerals.

Lines of force

The concept of lines of force is well known to every one who has handled a bar magnet or a compass needle. In the laboratory it can be demonstrated by sprinkling iron filings at random over a thin glass sheet placed on a bar magnet. The curves along which the filings orient themselves are referred to as "lines of force". Alternatively, the lines of force can be traced by following the directions assumed by a small compass needle when placed at various points around a magnet (Fig.85).

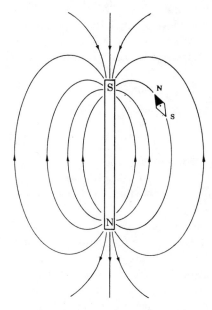

Fig.85.Lines of force due to a bar magnet. The lines of force spread out from the north pole of the magnet and converge to the south pole.

The points in which the lines of force seem to converge at the ends of a long, thin magnet are known as the "poles" of the magnet; these poles are actually located slightly inside the magnet. By convention, the "north-seeking" pole corresponding to that at the north end of a compass needle is called the *positive* pole, and the "south-seeking" pole is referred to as the *negative* pole. The lines of force are directed outward from a positive (i.e. north) pole and inward to a negative (i.e. south) pole.

Magnetic fields

The interactions between magnets are calculated by assuming that each magnet has a positive and a negative pole situated near the ends of the magnet and that the force between magnetic poles is given by a law analogous to Coulomb's law for electric charges, or to Newton's law of gravitation for masses.

The force acting between two magnetic poles of strength m and m' separated by a distance, r, is then given by:

$$F_0 = \frac{\mu_0}{4\pi} \frac{1}{\mu_R} \frac{m\,m'}{r^2} \qquad (4.1)$$

Here F_0 is the mechanical force, in newtons; r in metres; μ_0 (regarded as a universal constant) is the permeability of vacuum and has a numerical value of $4\pi \times 10^{-7}$ in SI-units; μ_R is the relative permeability of the medium, a dimensionless quantity that will be defined later.

Following the I.A.G.A.* recommendation in 1973 to adopt the International System of Units in geomagnetism, the unit for magnetic pole strength m is ampere × metre (A m). The unit for μ_0 then becomes newton/ampere² (N/A²) to make eq. 4.1 homogeneous, but other compounded units for μ_0 are in common use and they will be mentioned later.

The magnetic field, B, due to a pole of strength m at a point P, distance r, can be expressed in terms of the force of attraction per unit pole at P. It follows from eq. 4.1 that the B-field at P is:

$$B = \frac{F}{m'} = \frac{\mu_0}{4\pi} \frac{1}{\mu_R} \frac{m}{r^2} \qquad (4.2)$$

The SI-unit for the B-field is tesla, which from eq. 4.2 can be expressed as N/A m.

The magnetic field can also be described in terms of the

*International Association for Geomagnetism and Aeronomy.

field of force that surrounds electric currents. An electric current produces at every point in its "vicinity" a magnetizing field, H. The value of H can be calculated from the law of Biot and Savart, provided the electric circuit is given. As a result of the magnetizing field H, the current-carrying circuit is surrounded by a magnetic flux (lines of magnetic induction). The density of this flux (flux per m² perpendicular to the direction of the flux) is referred to as the magnetic field or the B-field, whereas the causative magnetizing field, H, is referred to as the H-field. An electric current is thus surrounded by two fields: the H-field and the B-field.

For almost all media, the flux density (the B-field) is found to be proportional to the magnetizing field (the H-field), to which it is due. The relationship between the two is given by:

$$B = \mu H \qquad (4.3)$$

where μ is the absolute permeability of the medium on which H is acting. It will be shown later that $\mu = \mu_0 \, \mu_R$.

The unit of B-field follows from Lenz's law of electromagnetic induction. This law states that the time rate of the change of magnetic flux $(-dN/dt)$ associated with a circuit is numerically equal to the electromotive force (in volt) induced in it. The magnetic flux, N, can therefore be measured in volt × second, and the flux density, B, in V s/m² or Wb/m² since 1 volt × second is called 1 weber.

The unit for the B-field is Wb/m², which in SI-units has the name tesla (T). The dimensions of B are: V s/m² (cf. eq. 4.2). In the electromagnetic centimetre-gramme-second (c.g.s.) system of units, the unit for B is gauss (G), which equals 10^{-4} Wb/m² or 10^{-4} T. For geophysical purposes, a sub-unit, the gamma (γ), which equals 10^{-5} G, appears to be more convenient. It follows that $1 \gamma = 10^{-9}$ T (or nanotesla). The total magnetic field of the earth in polar regions has a flux density (on the surface) of the order of 60,000 γ.

The unit for the *H*-field follows from Biot and Savart's law. At the centre of a loop of wire (of radius *r*) carrying a current *i* the *H*-field is:

$$H = i/2r \qquad (4.4)$$

The SI-unit for *H* is A/m. In the electromagnetic c.g.s. system, *i* is measured in absolute amperes (1 ab. amp = 10 "ordinary" amp), *r* in cm, and the magnetizing field *H* in oersteds (Oe). Numerically 1 A/m equals $4\pi \times 10^{-3}$ Oe.

The unit oersted is sometimes used in geophysical literature for expressing the magnetic field in the sense of flux density. This usage must be regarded as incorrect, irrespective of whether the flux density, *B*, is measured in vacuum, air or within a magnetic substance. However, in non-magnetic media, such as air or water, gauss and oersted are numerically equal, and any physical distinction between them is of little practical importance.

Intensity of magnetization

Let us consider a bar magnet of length *l* and cross-sectional area *A*, assumed to be magnetized uniformly in the direction of *l*. Referring to Fig.86 the bar magnet can be considered as a series of small elementary magnets (dipoles) oriented along its axis. The magnetic intensity due to the individual north and south poles of the elementary magnets cancel one another

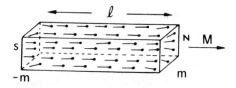

Fig.86.Schematic representation of a uniformly magnetized bar as a series of small elementary magnets (dipoles). The intensity (or strength) of magnetization can be expressed in terms of pole strength, *m*, or moment, *M*, as explained in the text.

except at the end faces. Thus, in effect the bar magnet will have a surface concentration of free positive (N) and negative (S) poles of a total strength, say m, at each end face. The greater the magnetization of the body is, the greater would be the surface concentration of free poles. The intensity of magnetization is a measure of the pole strength per unit area at the end faces, and can be expressed as:

$$J = m/A \left[\frac{\text{amp. metre}}{\text{metre}^2} \right] \qquad (4.5)$$

The SI-unit for J is A/m.

An alternative way of defining J is in terms of the magnetic moment, M. The bar's magnetic moment is expressed as the product:

$$M = ml \ [\text{ amp. metre}^2 \ ] \qquad (4.6)$$

If V is the volume of the magnet, then using eqs. 4.5 and 4.6 J can be expressed as:

$$J = ml/V = M/V \qquad (4.7)$$

Hence the intensity of magnetization, J, at any point within a uniformly magnetized body can be defined as the magnetic moment per unit volume.

Being directly measurable, the moment M is the most important parameter of a magnetized body. The SI-unit for magnetic moment is A m². This agrees with the dipole moment of a small, plane electric current loop, which is defined as the product of the current and loop area (A m²).

For a uniformly magnetized body, the magnitude and direction of J are the same throughout the body. In actual fact even if the magnetic material is evenly distributed throughout the body, the assumption of uniform magnetization is only approximately correct except for homogeneous bodies of spherical or

ellipsoidal shape. The intensity of magnetization, J, is a fundamental quantity for describing the magnetic state of the body.

Dipole field

The concept of a magnetic dipole is basic for an understanding of the magnetic behaviour of matter ranging in dimensions
from small magnetic particles to the entire earth. Mathematically we may consider a dipole to consist of two magnetic poles
of strength $+m$ and $-m$, whose physical size and separation
are infinitely small, but whose moment, $M = ml$, is nevertheless finite. Thus, a dipole represents an idealized elementary
magnet.

We shall now derive an expression for the magnetic field due
to a dipole (an elementary magnet) at a point, P, separated by
a distance, r, from the centre of the magnet (Fig.87). This
simple calculation helps to understand quantitatively the magnetic effects produced by magnetic bodies. By analogy with the

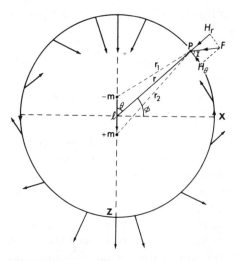

Fig.87.Magnetic field due to a dipole. To a first approximation, the earth's
magnetic field can be modelled by placing a dipole of equivalent moment
at the centre of the earth.

gravitational potential due to a point mass (see p. 91), the magnetic potential, W, at P is the sum of the contributions of the positive and negative poles of the magnet:

$$W = \frac{\mu_0}{4\pi}\left(\frac{m}{r_1} - \frac{m}{r_2}\right) = \frac{\mu_0}{4\pi} M \frac{\delta(1/r)}{l} \qquad (4.8)$$

where $\delta(1/r)$ represents the small difference $(1/r_1 - 1/r_2)$. Further, if l is infinitely small, $\delta(1/r)/l$ is the gradient of $(1/r)$ in the direction of l. Now if we choose a coordinate system such that l is in a direction z, and $r = \sqrt{x^2 + y^2 + z^2}$, then the expression for W can be written as:

$$W = \frac{\mu_0}{4\pi} M \frac{\delta}{\delta z}(1/r) = \frac{\mu_0}{4\pi}\frac{M}{r^2}\cos\theta \qquad (4.9)$$

The negative derivative of the potential yields the magnetic field intensity in the corresponding direction. Accordingly, the magnetic field (B-field) in the direction of r at point P is:

$$H_r = -\frac{\delta W}{\delta r} = \frac{\mu_0}{4\pi}\frac{2M}{r^3}\cos\theta \qquad (4.10)$$

The perpendicular component, H_θ (normal to r), at P is:

$$H_\theta = -\frac{1}{r}\frac{\partial W}{\partial\theta} = \frac{\mu_0}{4\pi}\frac{M}{r^3}\sin\theta \qquad (4.11)$$

Denoting the two field components at P by Z_0 ($= H_r$) and H_0 ($= H_\theta$), the resultant field F ($= \sqrt{Z_0^2 + H_0^2}$) is inclined to H_0 at an angle I, which is given by:

$$\tan I = Z_0/H_0 = H_r/H_\theta = 2\cot\theta = 2\tan\phi \qquad (4.12)$$

The above equations for a dipole field are fundamentally important in geomagnetism. If we made a sphere of wood with a very small bar magnet at its centre, the magnetic field produced at a point P lying on the surface of the sphere would have a direction as that shown in Fig.87. From the potential theory it can be shown that the external field of a uniformly magnetized sphere is identical to that of a dipole of equivalent moment placed at its centre. That the earth's field resembles the field of a magnetized sphere was first recognized by William Gilbert. We shall return to a discussion of the geomagnetic field in a later section.

Magnetic susceptibility and permeability

When a magnetizeable body is subjected to an external magnetic field, H, it acquires a certain degree of magnetization which is lost when the applied field H is removed. Such a magnetization, J_i, is said to be induced by the applied magnetizing field H.

However, some materials (e.g. iron, cobalt, nickel) and many rocks (notably basaltic lavas) exhibit a strong magnetic action even in the absence of an applied field. This phenomenon, as will be explained on p. 173, is attributed to their permanent or remanent magnetization.

For the present we shall restrict our discussion only to those materials for which the induced magnetization, J_i, is parallel and proportional to the applied field H. For such materials we may write a simple relationship:

$$J_i = kH \qquad (4..3)$$

The factor k is called the magnetic susceptibility and is a characteristic constant for a magnetizable material.

In SI-units, k is "dimensionless", since J_i and H are both measured in the same units (A/m). In the unrationalized version of the electromagnetic c.g.s. system, J_i is measured in absolute ampere per centimetre but H in units of "$1/4\pi$

absolute ampere per centimetre" (= 1 Oe.). The magnitude of k expressed in SI-units is 4π times that expressed in the un-rationalized electromagnetic c.g.s. system.

The flux density caused by a magnetizing field H in a region of vacuum (or air) is simply $B_0 = \mu_0 H$. When a magnetizable body is brought into this region the induced magnetization J_i ($= kH$) creates an additional flux density in the region now occupied by the body. The total flux density within the body therefore becomes:

$$B = \mu_0 H + \mu_0 k H = (1+k)\, \mu_0 H = \mu_R \mu_0 H \quad (4.14)$$

The factor $(1 + k)$, denoted by μ_R, is called the relative permeability. Like k, it is dimensionless. The quantity $\mu_0 (1 + k)$, representing the product $\mu_0 \mu_R$, is known as the absolute permeability μ. Its dimensions are the same as for μ_0.

It is important to note that for most magnetic minerals the linear relation between J_i and H (and also between B and H) is only approximately valid. As will be shown later, the susceptibility of ferromagnetic minerals, in general, depends on the strength of the applied field H and on the magnetic history of the minerals.

Diamagnetism, paramagnetism and ferromagnetism

A substance is called *diamagnetic* if the magnetic susceptibility, k, is negative; in the case of *paramagnetic* and *ferromagnetic* substances k is positive. The origin of these basic types of magnetism can be explained from the atomic structure of the substance.

In accordance with the laws of electromagnetism, all atoms have a magnetic moment due to the orbital motion and spin of their electrons. In most materials the magnetic moments of adjoining atoms are randomly oriented in the absence of an external field so that a specimen has no resultant magnetism. When a magnetic field is applied a magnetization is induced. In diamagnetic materials the induced magnetization, J_i, is in opposite direction to the applied field, H_a, which makes the susceptibility negative. All substances are diamagnetic, but many substances have other superimposed effects that obscure this weak magnetism. Many common minerals, such as quartz, feldspar, gypsum and rock salt are dominantly diamagnetic and have small negative values of susceptibility ($k \sim 10^{-5}$ SI).

Paramagnetism is dominant in atoms with an odd number of electrons, and it is mainly due to the unbalanced spin magnetic moments. Normally the moments are distributed randomly, but in the presence of a magnetic field they tend to line up in the direction of the field, the tendency being resisted by thermal agitation. The susceptibility of paramagnetic substances is positive and decreases inversely with the absolute temperature (Curie-Weiss law). Paramagnetism is generally ten or more times stronger than diamagnetism. Paramagnetic minerals that are frequently found in natural rocks are, for example, pyroxenes, olivines, garnets, biotites and amphiboles.

In certain other materials, broadly classified as ferromagnetics, the susceptibility is several orders of magnitude greater, sometimes more than 10 SI. It is positive (like that of paramagnetics), and is dependent on temperature and on the strength of the applied field. In these materials the spin moments of unpaired electrons are magnetically coupled between the neighbouring atoms, and this interaction results in a strong "spontaneous magnetization" even in the absence of an external field. The other remarkable property is their ability to retain the alignment imparted by an applied field after it has been removed. The regions within which the interaction and alignment of moments extend have dimensions of the order of 10^{-6} m and are called the magnetic domains. The alignment of atomic moments within a domain may follow one of the patterns shown in Fig. 88. In ferromagnetic materials, such as Fe, Co and Ni, the moments are parallel. In ferrimagnetic substances, of which magnetite (Fe_3O_4) is an example, the moments are arranged in two unequal sublattices, resulting in a net residual moment. The spontaneous magnetization of ferromagnetic and ferri-

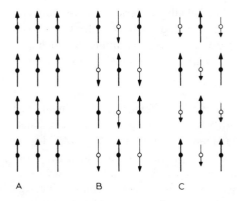

Fig.88. Diagrammatic representation of atomic moments in ferromagnetic (A), antiferromagnetic (B), and ferrimagnetic (C) crystals. (After Nagata, 1961.)

magnetic substances decreases with increasing temperature, and disappears at the "Curie temperature" T_C, which is about 580°C for magnetite. In antiferromagnetic substances the moments are arranged in two equal antiparallel sublattices so that normally there is no net moment. However, as the temperature is raised first one sublattice becomes disordered causing a relative increase of magnetic susceptibility. As the temperature is increased further, the susceptibility drops and the ordering of moments is lost at a critical temperature T_N (the Néel temperature) above which the crystal is paramagnetic. Certain antiferromagnetic minerals (such as haematite, $\alpha\text{-Fe}_2O_3$) usually exhibit a small net spontaneous magnetization due to the imperfect alignment in the sublattices or due to some other systematic defect. Such a weak ferromagnetism associated with imperfect antiferromagnetism is called "parasitic ferromagnetism".

Most of the naturally occurring magnetic minerals are either ferrimagnetic or imperfect antiferromagnetic in their behaviour, although true ferromagnetism is common in extra-terrestrial rocks, meteorites and lunar samples which contain large amounts of iron and nickel alloys. Henceforth we shall use the term ferromagnetics in its broad sense in order to include the above mentioned subgroups of magnetic minerals which are mainly responsible for the bulk magnetism of rocks.

Magnetic hysteresis

Hysteresis is a characteristic of àll ferromagnetics including ferrimagnetic and imperfect antiferromagnetic minerals. This can be explained by the theory of domains (Fig.89A). In a field-free space ($H_{ex}=0$), the spontaneous magnetization of the domains in a crystal is arranged in a pattern related to the crystal axes, but with their directions of magnetism eliminating each other. With the application of an external field the domain walls move fairly easily, allowing more domains to be magnetized in the direction of H_{ex}. When H_{ex} is low, this process is reversible and the domain walls spring back into place when the field is removed. However, when H_{ex} is increased,

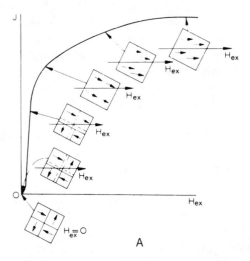

Fig.89.A. Magnetization process of a ferromagnetic substance according to the domain theory. Note the orderly alignment of domains with the increase in strength of the applied field H_{ex}.

one domain after another tends to swing in the direction of H_{ex}, the adjustments occurring with discrete jumps (called Barkhausen jumps) which are irreversible. At the saturation point the alignment of domains is complete and the magnetization, J, is maximum. When the field is removed, some of the domain walls remain in their newly acquired positions and the remaining magnetization corresponding to $H_{ex}=0$ is called the "remanent magnetization" (or simply the remanence). The above theory is valid for multi-domain grains (> 10 μm) which are normally observed in rocks. The magnetic behaviour of a single-domain particle is different, since it is not big enough to contain walls. However, the integrated effect of a random assemblage of single-domain grains is more or less similar to that of a multi-domain grain.

A hysteresis curve of cyclic magnetization in a ferromagnetic specimen is illustrated in Fig.89B. The specimen is magnetized to saturation, J_s, in a field, H_s. Then, as H_{ex} is reduced to zero, some residual magnetization (or remanent magnetization), J_r, is retained by the specimen. To eliminate this rema-

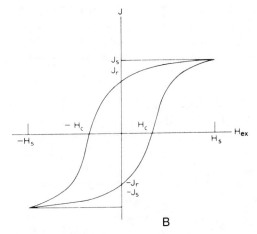

Fig.89.B. Hysteresis curve showing a cycle of magnetization of a ferromagnetic substance. (After Irving, 1964.)

nent magnetization a negative field, $-H_c$, called the "coercive force", has to be applied. The coercivity, H_c, is a measure of the hardness of the remanent magnetization. Generally the coercivity spectrum is in the range $10-10^3$ Oe (1 Oe $= 10^3/4\pi$ A/m), depending on the constituent magnetic minerals and on their grain size. For finely divided haematite grains, which are commonly found in red sediments, H_c is of the order of a few thousand oersteds.

Because of the phenomenon of hysteresis, the bulk susceptibility of a rock specimen is dependent on the state of magnetization, on J_s of the minerals contained, and to some extent also on non-intrinsic parameters like grain size, internal stress, etc. whose effect is roughly represented by the coercive force, H_c. As a general rule, the larger a grain is the more domains it contains and the easier it is to magnetize, that results in a higher susceptibility. Small grains are magnetically hard and thus tend to have a low susceptibility and a relatively high coercivity.

Common magnetic minerals

Practically all the constituents which give a high magnetization to rocks are ferrimagnetic minerals (including parasitic ferromagnetics). These minerals can be divided into two geochemical groups: (1) the iron–titanium–oxygen group, and (2) the iron–sulpher group.

The iron–titanium–oxygen group. The geochemistry and magnetic properties of the iron–titanium oxides can best be represented in the ternary system $FeO-Fe_2O_3-TiO_2$. In the ternary composition diagram, illustrated in Fig.90, the straight lines connecting Fe_2O_3 and $FeTiO_3$, and Fe_3O_4 and Fe_2TiO_4 represent the important solid-solution series comprising the majority of magnetic minerals in rocks. These series are respectively designated as α and β series.

The members of the α series range in composition from haematite (α-Fe_2O_3) to ilmenite ($FeTiO_3$). They have a rhom-

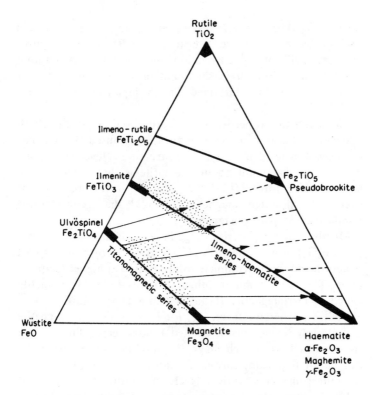

Fig.90.Ternary diagram of the composition of common iron titanium oxides showing the principal solid-solution series (heavy solid lines). The direction of increasing oxidation with a constant Fe/Ti ratio is shown arrowed. (After Tarling, 1971.)

bohedral structure. The magnetization of haematite is antiferromagnetic with a super-imposed parasitic ferromagnetism, whereas ilmenite is characteristically antiferromagnetic. The spontaneous magnetization, J_s, of haematite and ilmenite is very weak compared to that of magnetite. The chemical composition of the solid-solution series can be expressed as xFeTiO$_3$·$(1-x)$Fe$_2$O$_3$. The Curie (or Néel) temperature changes almost linearly with the increase of x from $675°$ for α-Fe$_2$O$_3$ to $-215°$C for $FeTiO_3$. The magnetic properties of the solid solution are complicated and sensitive to heat treatment, which can cause "self-reversal" of magnetization (see p.199). A curious example of such self-reversal found in a naturally occurring rock is that of the Haruna dacite in Japan (Nagata, 1961, p.176).

The magnetic properties of haematite itself are still not well understood because of the complications caused by spin imbalances and local defects in the structure. Nevertheless, the most interesting point about haematite is that it has a very high coercive force ($H_c > 10^3$Oe). It is for this reason that haematite, although magnetically weak has considerable significance in palaeomagnetism.

Members of the β series range in composition from magnetite (Fe_3O_4) to ulvöspinel (Fe_2TiO_4). Both are of inverse spinel structure and form a complete solid solution at higher temperatures. The most common mineral in this series is magnetite which is ferrimagnetic. It has a Curie temperature (T_C) of 580°C. The coercive force, H_c, for large grains of magnetite is quite low (some tens of oersteds); however, in fine particle sizes it can have a high magnetic stability so that magnetite can be an important carrier of stable remanence. Ulvöspinel is antiferromagnetic with a Néel temperature (T_N) of about -150°C; therefore it is paramagnetic at room temperature. The T_C and J_s of the solid-solution series decrease linearly as the composition moves from pure magnetite to ulvöspinel. A considerable amount of substitution of other bivalent and trivalent ions can occur in the spinel structure; Mg, Al, Cr, Mn are common in natural phases, thereby reducing J_s and T_C of the substituted magnetites.

Intermediate between the α and β series are the γ series of metastable iron–titanium oxides with a spinel (defect) structure. The end-member γ-Fe_2O_3 (maghemite) is ferrimagnetic with an unmeasured T_C. This is because pure maghemite is unstable at temperatures over 300°C when it converts irreversibly to haematite, although some impurities may help to stabilize its structure. Maghemite is present in many different rock types. In lava flows, it commonly forms at low temperatures of 200–300°C during the cooling process as the residual fluid reacts with the magnetite which, after being formed at higher temperatures, can be oxidized to some extent. It can also form in sedimentary environments by the low-temperature oxidation of small particles of magnetite in the presence of moisture. Apparently, there can very often be intermediate compositions, which are called titanomaghemites.

The iron–sulphur group. In this group the most important mineral is pyrrhotite (FeS_{1+x}, $0<x<0.15$). It is a ferrimagnetic mineral which is occasionally found in natural rocks. The standard composition of ferrimagnetic pyrrhotite is Fe_7S_8, which has a Curie point of about 320°C. Again, magnetic properties are complicated and sensitive to composition and heat treatment. For $x < 0.1$, the mineral is antiferromagnetic. When $x = 1$, we have the mineral pyrite, which has a cubic structure and is paramagnetic. There has been very little investigation of the magnetic properties of other sulphide minerals.

A group of minerals referred to as hydrous iron oxides (usually called limonites) is commonly found in weathered rocks. Goethite (α-FeOOH) is the oxyhydroxide corresponding to haematite, whereas lepidocrocite (γ-FeOOH) is the oxyhydroxide corresponding to maghemite. These minerals are thermally unstable and are therefore only of limited importance. Goethite commonly occurs in iron ore bodies and, along with haematite and magnetite, in weathered rocks. When heated above 120°C and cooled in an ambient field, it acquires a stable remanence with similar properties to those of haematite.

A more detailed account of the magnetic characteristics of rock-forming minerals can be found in a treatise by Nagata (1961); a good summary is given by Strangway (1970). Theoretical aspects of rock magnetism have been dealt with by Néel (1955), Verhoogen (1959), and Stacey and Banerjee (1974).

THE EARTH'S MAGNETIC FIELD

The fact that the earth is a huge magnet has been known for many centuries, and its magnetic field direction as indicated by a compass needle has remained most useful in navigation. The man who did most to found the science of geomagnetism was William Gilbert, a physician to Queen Elizabeth I. His discoveries described in the book *De Magnete* (published in 1600) gave the first quantitative account of the earth's magnetic field and showed that it resembled that of a magnetized sphere. Since then, the study of the earth's magnetic field continued steadily, and the knowledge of its behaviour and origin increased greatly. To understand better the significance of geomagnetic anomalies and rock magnetism as applied to various geological problems, we briefly summarize here the present knowledge of the earth's magnetic field.

The geomagnetic elements and poles

At any point on the earth's surface a magnetic needle, if free to move about its centre of gravity, will orient itself in a position determined by the direction of the geomagnetic field at that point. The geomagnetic field, F, is a vector quantity which requires the specification of three "elements" for a complete statement of its magnitude and direction at any point. There are alternative ways of choosing the set of three elements. The most common combination comprises the vertical component, Z, the horizontal component, H, and the declination, D, which is the angle between the direction of the horizontal component (i.e. the magnetic north) and the true or geographical north. An alternative set of elements is the total field intensity, F, its inclination, I, with respect to the horizontal, and the declination, D. Occasionally the field components are directly referred to geographical coordinates, north, X, east, Y, and vertically down, Z.

The quantities X, Y, Z, D, I, H, and F are known as geomagnetic elements, and their interrelationship can be derived from the diagram shown in Fig.91. Data can be converted from one set of elements to another by making use of the following relationships:

$$F = \sqrt{X^2 + Y^2 + Z^2} = \sqrt{H^2 + Z^2}; \quad \tan I = Z/H$$

$$X = H \cos D; \quad Y = H \sin D; \quad Z = F \sin I$$

$$(4.15)$$

The vertical plane through F and H is called the local magnetic meridian, whereas the plane XZ is the geographical meridian.

From about 1600 onwards, measurements of the geomagnetic field were regularly made at several observatories; although

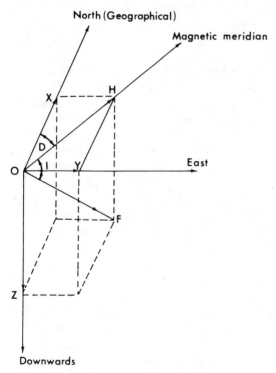

Fig.91. Main elements of the geomagnetic field. D and I are the declination and inclination, respectively, of the total field vector F.

most continuous records of the geomagnetic elements on world maps are available only for this century. The overall magnetic pattern as evident from these maps strongly suggests its similarity to the field which would be produced at the earth's surface if a powerful dipole magnet were placed at the earth's centre. From the analysis of geomagnetic data compiled on world maps, the magnetic dipole that best fits the actual field of the earth is found to have a moment of approximately 8×10^{22} A m^2 (8×10^{25} e.m.u.), with an axis inclination of about $11 \cdot 5°$ to the geographical axis (Fig.92). This corresponds to a horizontal intensity, H, of approximately 30,000 γ at the magnetic equator and a vertical field intensity, Z, of approximately 60,000 γ at each magnetic pole of the earth. The total intensity, F, of the geomagnetic field over the earth's surface is shown in Fig. 93.

The definition of magnetic poles requires a little caution. The axis of the best fitting central dipole intersects the earth's surface at two points which are referred to as "geomagnetic poles". The geographical coordinates specifying their locations are $78 \cdot 5°$N, $70°$W (in northwest Greenland) and $78 \cdot 5°$S, $110°$E (in Antarctica). On the other hand, the two principal points on the earth's surface where the magnetic field is directed vertically downwards ($I = 90°$, $H = 0$) are explicitly called the "dip poles". The magnetic dip poles (1965) are located approximately at $76°$N, $101°$W; $66°$S, $141°$E. They are not antipodal but correspond closely to an inclined, eccentric dipole situated some 400 km away from the centre of the earth.

The non-dipole field

Although the central dipole field (inclined at $11 \cdot 5°$ to the geographical axis) approximates the earth's actual field, there are considerable deviations between the two. When the best fitting dipole field is subtracted from the observed field, the difference obtained is called the non-dipole field or the "geomagnetic anomaly". Fig.94 shows the vertical component of the non-dipole field over the earth's surface. The important

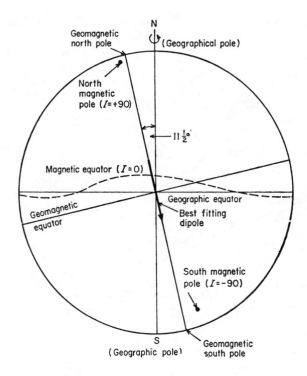

Fig.92.Illustration showing the distinction between the magnetic, geomagnetic, and geographic poles. (After McElhinny, 1973.) Actually, it is
the earth's *south magnetic pole* that is situated around the geographical
north and that attracts the north-pointing end of a compass needle. By
convention, however, the magnetic pole ($I = +90°$) around the geographical
north is referred to as the earth's *north magnetic pole*, and that around
the geographical south is referred to as the earth's *south magnetic pole*.

features of this map are some large-scale (regional) anomalies
extending over several thousands of kilometres, with amplitudes
reaching to 16,000 γ (roughly about 25% of the earth's total
field). These large-scale anomalies do not show any obvious
relation to geography or geology, and are almost certainly due
to deep sources within the earth. They are believed to originate
from regions near the core–mantle boundary and to be related

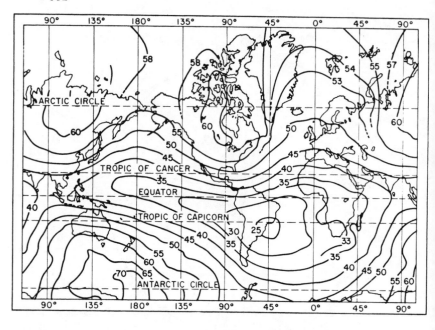

Fig.93.The total intensity of the earth's magnetic field expressed in units
of 1000 γ. (Source: U.S. Geol. Survey No. 1703.)

to certain cells of turbulent fluid motion at this boundary.

There are also several local anomalies of non-dipole field
that cannot be shown on a map of this size; the most remarkable one is at Kursk, south of Moscow, which has an amplitude
of about 190,000 γ. The origin of these anomalies is due to
local magnetization contrasts (which may be induced, or remanent, or mixed) in the crustal rocks, which are magnetic down
to a depth of about 20–25 km where the Curie-point isotherm
is reached. Presumably, therefore, there are no appreciable
magnetic sources between the Curie-point isotherm and the
earth's core.

Diurnal variation

Continuous magnetic records of any observatory show that

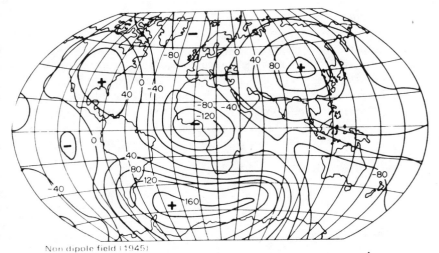

Non dipole field (1945)

Fig.94.The vertical component of the non-dipole field (or the "geomagnetic anomaly") expressed in units of 100 γ. (After Tarling, 1971.)

the earth's magnetic field is not constant, but is gradually changing with time both in intensity and direction. Short-period changes in the intensity of the field follow a daily cycle and show local variations of some tens of gammas during a normal "quiet" day. On the other hand, the "disturbed-day" variations are irregular and extreme in magnitude, amounting to several hundreds of gammas within an hour or so. They are associated with magnetic storms, which are related in some way with the increased solar activity during sun-spot cycles. Fig.95 shows the record of a base magnetometer reading registered during a magnetically stormy day in Nûgssuaq, West Greenland. Such storms usually last for several days.

Diurnal variations have considerable practical significance in magnetic surveying for geological mapping and mineral prospecting (see p. 211). Of course, during the "disturbed" days magnetic surveying operations have to be discontinued, since there is no satisfactory way of allowing for their unpredictable effect on magnetic field data.

Secular variation

Long-term changes in the geomagnetic field, which are progressive over decades or centuries, are known as secular variations. They are immediately apparent from the yearly averages of the values of geomagnetic elements recorded by magnetic observatories all over the world. The longest records of secular variation are from London and Paris which are sufficiently close together to show nearly the same pattern. Fig.96 shows

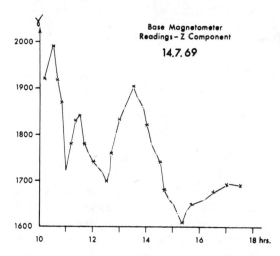

Fig.95. Diurnal variations in the vertical field intensity recorded at the base station in Nûgssuaq, West Greenland, during a magnetic storm on July 14, 1969.

the change in declination and inclination at London over the past four centuries. The curve suggests a cyclic variation with a period of about 500 years, but archaeomagnetic evidence, based on the study of direction of the remanent magnetism in archaeological objects of historically recorded ages, shows substantial departures from this periodicity.

Another interesting feature of the curve (Fig.96) is the clockwise rotation of the secular variation; it shows an apparent

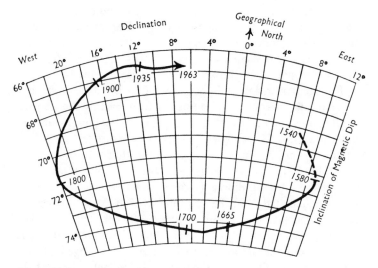

Fig.96.Observatory record of the variation in the inclination and declination of the geomagnetic field at London since the mid-16th century.(Based on the original plot of L.A. Bauer; redrawn from Parasnis, 1961.)

westward drift of the centre of secular variation by about 7° with respect to the zero declination. This corresponds roughly to the present dipole field position ; it implies that the dipole component of the field has remained nearly constant in orientation and that the secular variation is due primarily to the variability in the non-dipole field.

The westward drift is not only visible from the secular variation, but also from maps of the non-dipole field plotted at intervals of several decades. From a combined analysis of these variations, Bullard et al. (1950) found the average rate of westward drift to be 0.18° in longitude per year. At this rate, the pattern should go around the earth in about 2000 years. It has since been found that for many areas of the earth, the rate of westward drift is much smaller. However, it does appear, that in a relatively short time of 2000–5000 years, the outermost layer of the core (the region where the probable sources of the non-dipole field are located) could drift through $360°$ relative to the rest of the earth. When averaged over a time span of this order, the field at any point on the earth should very nearly correspond to that of a central dipole, a fact which is of vital importance in palaeomagnetism.

The secular variation is not limited to the declination and inclination. The magnetic moment of the earth's dipole itself is changing slightly from year to year, and therefore also its field intensity. The magnetic moment of a dipole can be calculated from the intensity of the field that it produces (see eqs.4.10, 4.11); in a similar way the magnetic moment of the earth's central dipole is determined. Since 1835, when Gauss calculated it for the first time $(8·5 \times 10^{22} \, A \, m^2)$, it has been steadily decreasing at a rate of about 5% per century. It was measured as $8·0 \times 10^{22} A m^2$ in 1960. This is a definite decrease which cannot be attributed to the use of improved instrumentation. At this rate, the main dipole magnetic field of the earth would disappear completely in another 2000 years, probably leaving nothing for the geomagnetists to work upon in the near future. Of course, we do not definitely know how long this trend will continue, but we do know from palaeomagnetic evidence (p.239) that a substantial decrease in the main dipole field could be an indication of the next change in polarity (or the so-called reversal) of the earth's field, which has been one of its most significant features in the geological past.

Origin of the main field

Although it explains local anomalies the magnetization of crustal rocks is too weak to account for the earth's main magnetic field. The temperatures in the mantle and core are too high (certainly above the Curie point of iron and nickel), for the material to retain ferromagnetism. Therefore, the earth's main field cannot be attributed to any permanent source of magnetization in the earth.

A satisfactory explanation of the cause of the earth's main

field must also be able to explain its rapid variations which have been recorded at magnetic field observatories for the past few centuries. As we noted, substantial changes in the direction of the field can occur within a few hundred years (see Fig.96). The time scale of secular variation has, in fact, provided sufficient evidence to prove that the earth's main field is unrelated to any geological processes occurring in the crust or mantle, which on the time scale are usually measured in millions, rather than thousands, of years. This places the origin of the field in the core, where the material is fluid and can respond rapidly to any forces imposed on it. Since the core is not magnetic, the source of the magnetic field must lie in the electric currents within the core.

The question then is: how can electric currents of the order of 10^9 amperes, which are necessary to account for the intensity of the earth's magnetic field, be generated and maintained in the core? In attempting to explain this, the simplest concept was put forth by E. C. Bullard and W. M. Elsasser, namely, that the rotation of the earth and some additional motions of the fluid within the electrically conducting but nonmagnetic core are sufficient for a self-exciting dynamo mechanism. The essence of their theory (see Fig.97) is that in the earth's core fluid motions act like the rotating disc in the disc dynamo. For further details of the dynamo theory, the reader is referred to the excellent articles by Elsasser (1958) and Bullard (1971).

To account for the periodic reversals of the field, a sophisticated model of two disc dynamos (coupled in series and rotating at different speeds) has been proposed. The extension of this model to the spherical case of the earth's liquid core is very complex and has not yet been carried out.

The precise nature and details of the forces that might drive the dynamo are still not known. In any event it seems certain that a mechanism is stirring up the core and causing fluid motions other than those due only to the earth's rotation. Three sources of energy have been suggested which might

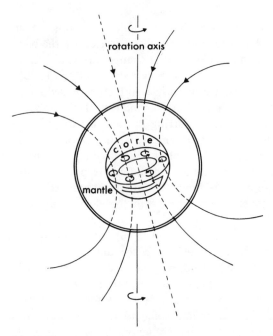

Fig.97.The earth's magnetic field is probably associated with an intense current loop (shown by the broad arrow) circulating in the core. The way in which this current is generated is not known in detail; it is presumably derived from local eddy currents produced in the liquid core by friction due to the earth's rotation.

keep the liquid core rotating and swirling sufficiently to account for the self-sustaining and self-reversing dynamo action.

(1) The inner solid core of the earth may still be enlarging and the heat radiated in ·the process of change from liquid to solid could provide the energy which maintains the fluid motions.

(2) The fluid material of the core may not be able to follow the precessional motion of the earth in its orbit and the resulting lag could cause a stirring effect by friction between the mantle and the outer core.

(3) The convective motion and churning of the fluid core

could be the result of radioactive heating, perhaps produced by pieces of solid mantle protruding into the core, or more probably, by the radioactivity of the solid inner core.

MAGNETIZATION OF ROCKS

The ferrimagnetic minerals (including the parasitic ferromagnetics) contained in rocks are normally in the form of fine grains dispersed throughout the matrix of paramagnetic or diamagnetic minerals. The bulk of rocks, therefore, shows the characteristics of ferrimagnetism. Magnetic characteristics of rocks depend partly on their mineral contents and partly on their history of formation. In general, both *induced* and *remanent* magnetization are to be considered when studying the bulk magnetization of rocks. In particular the use of remanent magnetization as an indicator of the fossil record of the geomagnetic field extending back to old geological periods, has added another dimension to the significance of the studies of magnetization in rocks. Moreover, geologic interpretations of magnetic measurements are critically dependent on the availability of techniques for distinguishing the various magnetization components. To have a better understanding and appreciation of the various applications of such studies, we first discuss briefly the different processes by which rocks can be magnetized.

Induced magnetization and bulk susceptibility

The magnetization produced in rocks by and approximately parallel to the locally acting external field (e.g. the geomagnetic field) is called the induced magnetization, which depends primarily on the susceptibility of a rock body.

When a rock specimen is put in a weak field H_{ex} (for example, that of the earth), the bulk induced magnetization, J_1, is expressed as:

$$J_1 = k_b H_{eff} \qquad (4.16)$$

$$H_{\text{eff}} = H_{\text{ex}} - H_{\text{D}} = H_{\text{ex}} - N_s J_1 \qquad (4.17)$$

where N_s denotes the demagnetization factor depending upon the shape of the specimen, k_b, the bulk magnetic susceptibility, and H_D the demagnetizing field within the specimen. The demagnetization effect due to shape can be understood by referring to Fig.98, where the distribution of "free" poles at the end faces of a body causes a field inside the body opposite in direction to the external inducing field.

In most of the rocks (except in magnetic ore bodies), J_1 is small ($\lesssim 1$ A/m or 10^{-3} e.m.u.) and therefore the demagnetizing field $N_s J_1$ is insignificant compared to H_{ex}. However, the effect of the demagnetizing field in individual ferrimagnetic

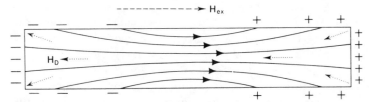

Fig.98.Schematic illustration of the demagnetization field within a body caused by the distribution of free poles. The demagnetization field, H_D (dotted arrows), acts in a direction opposite to the external inducing field, H_{ex}. Solid lines denote the direction of magnetization which, strictly speaking, is only uniform in homogeneous bodies of spherical or ellipsoidal shape.

grains is not negligible, because of the fairly large values of magnetization of individual grains.

As a first approximation, the magnetostatic interaction between the individual grains may be ignored provided the number of grains is small. With this simplification the bulk susceptibility of a rock specimen with an assemblage of grains can be expressed as:

$$k_b \simeq p\ \frac{k}{(1 + Nk)} \qquad \text{(for } p \ll 1) \qquad (4.18)$$

or :

$$k_b \simeq pk_a \qquad (4.19)$$

where p denotes the volume fraction of the magnetic material, k the isotropic intrinsic susceptibility, k_a the apparent magnetic susceptibility, and N the average demagnetization factor of the magnetic grains dispersed in the rock. When the shape of the magnetic grains is spherical, $N = 1/3$. For other shapes N ranges between zero and 1. Tables for N have been published by Stoner (1945) and Sharma (1966, 1971). Nagata (1961) found that k_b of many igneous rocks can be represented by eq. 4.18, with $k \sim 5$ SI and $N \sim 0.3$ for a typical assemblage of magnetite grains. However, deviations by a factor of 10 from the predicted value are quite common, reflecting mostly the effect of grain size on k. Some other empirical relationships have also been suggested to express k_b as a function of magnetite content in the rocks (e.g. Werner, 1945; Jahren, 1963).

From the above treatment it follows that the bulk susceptibility of rocks depends not only on the content of magnetic minerals, but also on their grain size and shape, their mode of dispersion, etc., and is extremely variable. Igneous and metamorphic rocks generally have higher susceptibilities than sedimentary rocks, but the range of variation is so great that it is not possible to identify rock types from the knowledge of the susceptibility and vice versa. It is, therefore, advisable to directly determine the susceptibility of rocks and minerals within the area of interest. The values listed in Table VI give a rough idea of the susceptibility of common rocks and rock-forming minerals.

The susceptibility can be measured either in field on rocks in situ using a handy instrument called the Kappameter*, or in the laboratory on collected samples. In-situ measurements are preferable wherever it is feasible to make them. For accurate measurement of low susceptibilities, special laboratory

*A commercial make of ABEM, Stockholm.

TABLE VI

Volume susceptibilities of rocks and minerals

Mineral or rock type	Susceptibility $(k \times 10^{-6}$ SI$)^*$	Remarks
Granite (without magnetite)	10–65	
Granite (with magnetite)	20–40,000	after Mooney and Bleifuss (1953)
Slates	0–1200	after Mooney and Bleifuss (1953)
Gabbro	800–76,000	after Mooney and Bleifuss (1953)
Basalt	500–120,000	after Bentz (1961)
Oceanic basalts	300–36,000	after Vacquier (1972, p.170)
Limestone (with magnetite)	10–25,000	rest of the data taken
Gneiss	0–3000	from a compilation by
Sandstone	35–950	Parasnis (1971)
Pyrite (ore)	100–5000	
Haematite (ore)	420–10,000	
Magnetite (ore)	$7 \times 10^4 – 14 \times 10^6$	
Magnetite (crystal)	150×10^6	
Serpentinite	3100–75,000	
Graphite (diamagnetic)	−80 to −200	
Quartz (diamagnetic)	−15	
Gypsum (diamagnetic)	−13	
Rocksalt (diamagnetic)	−10	
Ice (diamagnetic)	−9	

*To convert the above values to unrationalized electromagnetic c.g.s. units divide by 4π.

methods employing astatic magnetometers or a.c. bridge devices have been developed (for various designs, see Collinson et al., 1967). A versatile low-field susceptibilitymeter (Fig.99) has been designed by Radhakrishnamurthy and Likhite (1966) which also enables a visual display of the low-field hysteresis loop of a rock specimen. As susceptibility is strongly dependent on the field strength (see p.174, section *Magnetic hysteresis*), susceptibility data must be measured and reported under known field strengths. Since it is most usual in geophysical problems to be interested in the magnetic properties of rocks in the presence of the earth's field, a magnetizing

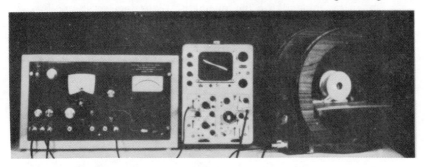

Fig.99. Susceptibilitymeter and hysteresis loop tracer. (Courtesy of the Rock Magnetism Group, T.I.F.R., Bombay.)

field of the order of 0.5 Oe(1 Oe $= 10^3/4\pi$ A/m) is commonly used.

Magnetic anisotropy and rock fabric

The magnetic susceptibility, k_b, of rocks is not always isotropic. It may be significantly anisotropic particularly in metamorphic and sedimentary rocks. Two types of anisotropy are to be considered.

When the shape of the magnetically isotropic grains in a rock is elongated along a special direction, k_b may become anisotropic. This is called the shape anisotropy and dominates in high-susceptibility cubic minerals such as magnetite. It occurs often in sedimentary and metamorphic rocks. The plane of maximum susceptibility generally lies in the bedding plane for sediments and in the plane of foliation for metamorphic rocks.

In some rocks, the major ferrimagnetic minerals themselves have a marked magnetocrystalline anisotropy depending on the direction of H_{ex} with respect to the crystal axes. The minerals of the ilmenite–haematite series and pyrrhotite are typical examples of crystalline anisotropy.

Since the magnitude of anisotropy for a grain of a given shape increases rapidly with increasing susceptibility, ferrimagnetic minerals of the magnetite–ulvöspinel series are by far the most important contributors to the anisotropy of rocks. When magnetic anisotropy is significant, J_1 is not parallel to H_{ex} (Fig.100) and therefore, k_b is usually represented in terms of a triaxial ellipsoid with axes k_{max}, k_{int} and k_{min}.

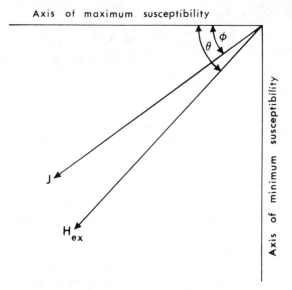

Fig.100. Deflection of magnetization, J, from the direction of the applied field, H_{ex}, due to anisotropy.

Attention was first drawn by Graham (1954) to the possibility of using susceptibility anisotropy in petrofabric studies. Since then the subject has been studied in more detail and the technique has now reached a stage where measurement of the susceptibility ellipsoid can be made rapidly; the main axes so determined can be used to define the plane of foliation (or bedding) and the megascopic lineation. The details of the various methods of measurement and analysis of the susceptibility anisotropy data are beyond the scope of this book. However, we shall briefly review some of the problems and results of magnetic fabric analysis in the three types of rocks.

(1) *Sedimentary rocks.* The method of estimating grain orientation by the measurement of the susceptibility anisotropy has proved useful in a number of studies. The two important sources of grain alignment are the action of gravity in producing a foliation parallel, or nearly parallel, to the plane of bedding, and the action of depositing water currents in producing a lineation in the direction of the flow. Fabrics containing a foliation and a single, flow-produced, lineation are very common, and the directional properties of such fabrics can be completely specified in terms of the $k_{max}-k_{int}$-plane (foliation) and the k_{max}-axis (lineation). One of the earliest studies was made on Swedish varves (Granar, 1958) which demonstrated k_{max} to lie in the horizontal plane, its magnitude being 10–20% more than that of k_{min}. An excellent review of the magnetic fabric of sediments, with some examples of palaeocurrent studies is given by Hamilton and Rees (1970). The method has also been used to study the fabric in till related to glacial movement (Fuller, 1962).

(2) *Metamorphic rocks.* Structural geologists have, for many years, used the fabric of metamorphic rocks to estimate stresses. The magnetic method is of great potential value to them especially in rocks which have no obvious lineations or foliations. A convincing demonstration of the value of the method is provided by the studies of Balsley and Buddington (1960) on granites and gneisses in the Adirondack mountains.

They found close correlations between visible and magnetic foliations and lineations and were also able to relate the magnetic fabric of rocks without visible structures to broad geological structural trends. Metamorphic rocks often have a complex fabric with multiple foliations or lineations and the method has to be applied with great care.

(3) *Igneous rocks*. It is difficult for the petrologists to study the microfabric of igneous rocks, even by tediously observing the rocks under a microscope. Recently attempts have been made to determine magnetically the fabric of lavas and igneous intrusives with the view to obtain information about their flow history and eventually their mode of formation. Examples of such studies are given by Khan (1962) and Symons (1967).

The knowledge of magnetic anisotropy is also important in the interpretation of magnetic anomalies, and, more specifically, in palaeomagnetic studies for testing the basic postulate of parallelism of the remanence to the magnetic field in which it was acquired.

Various methods for the laboratory measurement of magnetic anisotropy in rocks can be found in a collection of papers edited by Collinson et al. (1967).

Remanent magnetization in rocks

Most rocks constituting the upper part of the earth's crust have a remanent magnetization, J_r, in addition to J_i ($= k_b F$) induced by the present earth's field, F. The intensity of J_r is particularly large in igneous and thermally metamorphosed rocks, for which the Königsberger ratio Q ($= J_r/k_b F$) is often greater than 1 and sometimes exceeds even 100 (see Table VII, p.202). Hence the interpretation of magnetic anomalies, especially in areas where igneous rocks occur, must take into account the intensity and direction of J_r if a satisfactory geological picture is to be obtained.

The intensity of J_r depends on the amount and nature of the ferromagnetic minerals present, on the geomagnetic field strength at the time of origin of the remanence, and on the subsequent geological history of the rock body. The intensity of J_r has been found to range from 10^{-4} A/m (in SI) for weakly magnetized sediments to 10 A/m for basic igneous rocks. The majority of remanence measurements are now made with highly sensitive astatic and spinner magnetometers (for various designs, see Collinson et al., 1967). The remanence of moderate to strongly magnetized rocks can, however, be measured by a simple flux-gate device (Sharma, 1968) or by an ordinary field magnetometer (Parasnis, 1973, p.351).

In addition to the intensity, the direction of the remanence is of basic importance to the studies of the magnetic properties in rocks and also in the interpretation of geomagnetic anomalies. The direction of J_r, except in very recent rocks, may differ significantly, often by as much as $180°$, from that of the present geomagnetic field. The subject of palaeomagnetism, with its most exciting applications, is based on the study of the direction of J_r in rocks of all geological ages from different continental blocks and ocean floors.

During the past decades considerable progress has been made in understanding the various processes by which natural remanent magnetization (NRM) is acquired by rocks. The geophysically most important ones are the following:

(1) *Isothermal remanent magnetization (IRM)*. IRM is the remanence acquired by a rock at constant temperature, when an external field is applied for a short time and then removed as is done while displaying a magnetic hysteresis loop (p.173). It is of negligible intensity in weak magnetic fields, e.g. that of the earth. IRM is, therefore, important only in rock outcrops that have been exposed to local, intense fields, such as may be produced by lightning strokes (Graham, 1961). Fortunately, the magnetic effects brought about by a lightning stroke are very local, highly irregular in intensity of IRM, and therefore easily recognizable.

(2) *Viscous remanent magnetization (VRM)*. VRM is a cumulative IRM acquired by rocks after a long exposure in an ambient field similar to that of the earth. The increase in

intensity of VRM is generally a logarithmic function of time, and the hardness of VRM increases with the passage of geological time. Rocks vary considerably in their magnetic "viscosity" and the VRM acquired is inversely related to their coercivity, H_c. For example, magnetite ores, which are magnetically "softer" than basalts, develop in 70 days a VRM as large as about 20% of the TRM (Sholpo, 1967). In some cases their remanence can be considerably changed by storing them in the laboratory for a period of only a few weeks. Rocks whose NRM is subject to such large viscous effects are clearly unsuitable for any meaningful palaeomagnetic studies.

(3) *Thermoremanent magnetization (TRM)*. TRM is acquired by rocks during their cooling from the Curie point to room temperature in the presence of a magnetic field. This is the most important mechanism in understanding the relatively strong and stable NRM of many igneous rocks. Most of the TRM is usually acquired in a temperature interval of $100°$ or $150°C$ below the Curie point (T_c). In most rocks the TRM is faithfully parallel to the ambient field, H_a, and for low field strengths it is proportional to H_a.

Partial thermoremanent magnetization (PTRM) has the interesting property that the total TRM produced by cooling from T_C to room temperature T_0 turns out to be the sum of all PTRM's acquired in the intervals T_C-T_1, T_1-T_2 ..., T_n-T_0. Conversely, reheating to temperature $T_i < T_C$ and subsequent cooling in a zero field destroys only that part of the original TRM that was acquired below the temperature T_i. This remarkable property of PTRM is found to apply in baked pottery and many (but not all) lavas. In suitable cases, it can be used to determine the palaeointensity of the earth's field that produced their original NRM (Thellier and Thellier, 1959).

In weak fields, TRM is much stronger and stabler than the IRM acquired in the same field. This stability of TRM permits separation of the original TRM from any secondary magnetization (e.g. VRM) acquired later in many cases. The stability of TRM is roughly proportional to the coercivity and usually decreases with increasing grain size. A simple theory based on single-domain particles that accounts for many of the observed TRM properties has been developed by Néel (1955).

(4) *Self-reversed magnetization* (*SRM*). SRM is acquired in a direction opposite to that of the ambient field acting during the acquisition of NRM. As mentioned previously (p.176), self-reversal of NRM is known to occur in some α phases (ilmenite–haematite series); it may perhaps also be produced in β and γ phases by impurities and partial oxidation (Verhoogen, 1962), or by magnetostatic interaction between two constituents with different Curie temperatures (Néel, 1955; Uyeda, 1962; Bhimasankaram, 1964). Until now only a few artificial materials (e.g. lithium–chromium ferrite) and very few natural rocks containing ilmeno-haematites and pyrrhotites are known to show SRM in a repeatable and convincing manner. However, the subject demands a thorough study in order to exclude the possibility of any SRM before attributing all the reversed NRM of rocks to the polarity inversions of the geomagnetic field.

(5) *Depositional remanent magnetization* (*DRM*). Grains of magnetic minerals that carry a remanence (e.g. TRM) acquired earlier may become oriented by the earth's field when they settle in still water and become imbedded in sediments, thereby giving the consolidated sediments a depositional or detrital remanent magnetization. Fine-grained sediments tend to have a relatively stable DRM. Varved clays provide the best example (Granar, 1958). Mineralogical studies show that the main magnetic material in varves is magnetite in very fine particle sizes of only a few microns.

Extensive laboratory tests (King, 1955) show that certain effects known as "inclination error" and "bedding error" may cause the inclination of DRM to deviate considerably ($10°-25°$) from the actual direction of the ambient field.

(6) *Crystallization or chemical remanent magnetization (CRM).* CRM is acquired at the time of nucleation and growth or recrystallization of fine magnetic grains by certain chemical reactions (at temperatures far below their Curie points) in an ambient field. The mechanism appears to be very similar to that which produces TRM in single-domain particles (Kobayashi, 1959). The stability of CRM, with respect to thermal demagnetization or to an a.c. field demagnetization (see p.236), is very similar to that of TRM (Fig.101), although the intensity is not so great. It seems that many sedimentary and metamorphosed rocks in nature possess CRM. Some of the remanence of red sediments may be a post-depositional CRM acquired by dehydration and recrystallization of goethite to haematite, or of lepidocrocite to maghemite. Even in igneous rocks the iron oxide minerals may undergo a transition from one form to another during a slow exsolution or a low-temperature oxidation process and may

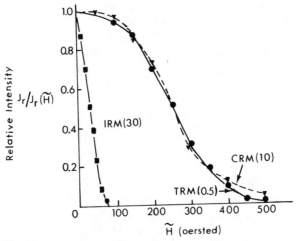

Fig.101.Alternating field demagnetization curves for various types of remanent magnetization in magnetite samples. TRM was acquired in a 0·5-Oe, CRM in a 10-Oe, and IRM in a 30-Oe field ($1 \text{ Oe} = 10^3/4\pi$ A/m). (After Kobayashi, 1959.)

then acquire a CRM.

(7) *Piezo-remanent magnetization (PRM)*. PRM is an additional remanent magnetization acquired by the application and release of mechanical stresses in an ambient field at a constant temperature (Nagata, 1961). Also a pre-existing remanence may be decreased by uniaxial compression in the direction of magnetization in much the same way as the susceptibility. Studies of PRM and the related phenomenon, magnetostriction, have recently acquired special significance, since the piezomagnetic effects associated with the critically stressed rocks prior to an earthquake could give very minute, but measurable, seismomagnetic anomalies of a few gammas (Stacey, 1969, p.123). Such effects may give an indication of an impending earthquake. In recent years several arrays of sensitive magnetometers have been installed along the San Andreas fault and at other sites in the western United States (Kovach and Breiner, 1967) to test the piezomagnetic effect of movements in fault zones.

The Königsberger ratio for rocks

In studying the magnetization of rocks, a numerical parameter, the Königsberger ratio Q, is often quoted. This is the ratio of the NRM present in a rock to the magnetization induced by the present earth's field at the sample location (i.e. $Q = J_r / k_b F$). Because of the relaxation processes of various types, which tend to weaken the original NRM with time, older rocks are usually found to have smaller Q values than younger rocks of the same type. The range of the Q values is very large however, occasionally exceeding the limits $0 \cdot 1 < Q < 100$.

Table VII gives the data for the ratio Q of different rock types compiled from several selected localities. It is apparent from the data that the Q ratio even for the same rock type may be extremely variable; there is an obvious need for much more data of this kind especially for crystalline basement rocks. Nevertheless, it might be useful to keep in mind some general-

TABLE VII

Königsberger ratio (Q) for some rock types

Rock type	Locality	(Q)	Reference
Oceanic basalts	EM-7 Mohole Northeast Pacific	15–105	Vacquier, 1972, p. 169 (mean $Q_n=40$)
Oceanic basalts	Mid-Atlantic Ridge	1–160	Vacquier, 1972, p. 170 (mean $Q_n=48$)
Sea-mount basalts	North Pacific	8–57	Vacquier (1972, p. 172)
Cainozoic basalts	Victoria, Australia	5	Irving and Green (1957)
Early Tertiary basalts	Disko, West Greenland	1–39	Sharma (1975)
Dolerite sills	North England	2–3·5	Creer et al. (1959)
Gabbro	Småland, Sweden	9·5	Parasnis (1972, p. 9)
Gabbro	Minnesota, U.S.A.	1–8	Jahren (1965)
Diabase dykes	Canadian shield	0·2–4	Strangway (1966)
Pilansberg dykes	South Africa	2–5	Gough (1956)
Granite plutons	California, U.S.A.	0·2–0·9	Currie et al. (1963)
Granodiorites	Nevada, U.S.A.	0·1–0·2	Allingham and Zietz (1962)
Granitic intrusives	Japan	0·1–0·5	Ito (1964)
Manganese ore	South India	1–5	Bhimasankaram and Rao (1958)
Magnetite ore	Sweden	1–10	Parasnis (1972, p. 9)
Sedimentary rocks (except some ores having CRM)		<0·1	Nagata (1961, p. 312)

izations made by Nagata (1961), which are as follows:

(1) In igneous and thermally metamorphosed rocks over continental areas, $Q \sim 1$ for well-crystallized rocks.

(2) For volcanic rocks $Q \sim 10$, and for many rapidly quenched basaltic rocks (e.g. oceanic basalts) $Q \sim 30\text{-}50$.

(3) In sedimentary and metamorphic rocks, Q usually is < 1, except in special cases of iron ores having CRM.

In general, the total magnetization of a rock is expressed as a vector sum of $J_i (= k_b F)$ and J_r. For rocks of Recent and late Tertiary age, the direction of J_r is found to be nearly parallel or antiparallel to that of the present geomagnetic field, F, the deviation being usually within approximately $25°$. In such cases the resultant magnetization can be simply expressed in terms of Q as:

$$J \simeq k_b F \pm J_r = k_b F (1 \pm Q) = k_E F \qquad (4.20)$$

where $k_E = k_b(1 \pm Q)$ can be considered as the "effective" magnetic susceptibility responsible for the total magnetization of a rock body. When k_E takes a negative value for basaltic and andesitic rocks, as it often does, then $Q > 1$ and J_r is approximately antiparallel to F. Therefore, the possibility of the presence of reversely magnetized rocks must be taken into consideration in the interpretation of negative magnetic anomalies.

MAGNETIC CHARACTER OF CONTINENTS AND OCEANS

From the thermal behaviour of ferromagnetic minerals, and in particular the decrease of the Curie point with increasing pressure (Nagata, 1961, p.325), it is quite apparent that the magnetization of rocks can extend only to certain limited depth (several tens of kilometres) bounded by the Curie-point isotherm for magnetite, which is by far the most common magnetic mineral in rocks. It has also been shown in a study of anomaly wavelengths measured over long magnetic profiles

(Alldredge et al., 1963) that there are probably no sources of magnetic anomaly in the mantle. The lower limit of the anomaly producing crustal blocks should, therefore, correspond to the Curie-point isothermal surface, which could of course be warped depending on local geothermal conditions. An upward warping will cause a relative decrease of the magnetic anomaly, whereas a downward warping will increase the anomaly. Pakiser and Zietz (1965) observed a distinct fall in the magnetic anomaly west of the Rocky Mountains, and they suggested that from the mountain front westward, the Curie isotherm is upwarped in the lower crust. It has been possible to map the Curie isotherm by magnetic surveys in several areas. Vacquier and Affleck (1941) made an analysis of 126 anomalies in the U.S.A., from which they estimated the Curie-point isotherm at a depth of about 20 km. From an analysis of 85 anomalies in the Canadian shield, Bhattacharya and Morley (1965) found the depths to the bottom of magnetized crust (interpreted to be Curie-point depths) to lie mostly between 17 and 24 km. Hahn and Zitzmann (1969) report the Curie-geotherm depth in northern Germany to be 20–25 km with an uncertainity of ± 5 km.

As to the magnetic character of crustal rocks, it follows from Tables VI and VII that both the susceptibility and the remanence of sedimentary rocks are too low to give them any appreciable degree of magnetization. Therefore, virtually all magnetic anomalies observed over continents and oceans should be attributed to igneous and metamorphic rocks which generally possess both types of magnetization (J_i and J_r).

Within the continents, the major areas of igneous and metamorphic rocks, either exposed or concealed beneath sedimentary deposits, are of Precambrian age. For these old rocks, J_r is very often smaller than J_i, and in general $Q < 1$. Even in those rock units where the intensity of J_r is comparable to that of J_i, the direction of J_r appears to vary irregularly to such an extent that its contribution to the magnetic anomaly produced by a large mass of rock is rather small. This is also shown by the fact that the magnetic anomaly maps, made over different continental areas in mid latitudes, nearly always display the general pattern of small negative anomalies (see Fig.111) on the north side (in the northern hemisphere) with approximately the expected ratio of amplitudes of positive and negative anomalies corresponding to the local inclination of the geomagnetic field. Of course, there are exceptions; certain intrusives (dyke swarms and sills) and almost all volcanics (e.g. Deccan traps covering the Indian shield) are known to have fairly strong magnitudes of J_r, and Q ratios >1. However, in general, the magnetic character of continents appears to be governed by variations in the susceptibility.

In contrast to the continents, the magnetic character of oceans appears to be predominantly controlled by the remanent magnetization of the basaltic layer in the oceanic crust. In recent years the magnetic properties of oceanic basalts (mostly from dredged samples) have been studied in great detail. These samples are often fine-grained and glassy due to the rapid cooling in seawater, which endows them with a lower susceptibility and a higher coercivity than similar continental basalts (Cox and Doell, 1962; Ade-Hall, 1964; Ozima et al., 1968). Although the mean intensity of their NRM is comparable to that of continental basalts, the Q ratio is distinctly larger than that of continental rocks (see Table VII). Thus, the magnetic anomalies observed over oceans could be interpreted in terms of the remanent magnetization only, in contrast to the induced magnetization of continental rocks.

The other distinct feature of the ocean floors is their linear pattern of magnetic anomalies. The linear anomaly trend off the west coast of California, first discovered and mapped in detail by Mason and Raff (1961), depicted a curious pattern of north–south oriented strip-like positive and negative magnetic anomalies (Fig.102). The offsets of the anomaly strips were originally considered to be indicative of large-scale horizontal or strike-slip movements associated with fracture zones, although they were later interpreted as transform

206

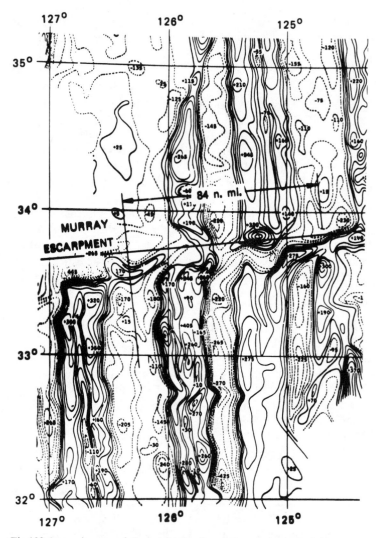

Fig.102.Anomaly map of the total intensity of the geomagnetic field, contoured in gammas. The offsets of the anomaly strips indicate a horizontal displacement of 153 km along the Murray fracture zone. (Based on Mason, 1958, fig. 2.)

faults (Chapter 8, p.363). Subsequent work has shown that strip-like magnetic anomalies are a characteristic feature of all oceans. Furthermore, it has been found that the pattern of anomalies is strikingly symmetrical about the crests of many ridges. A classic example is shown in Fig.103.

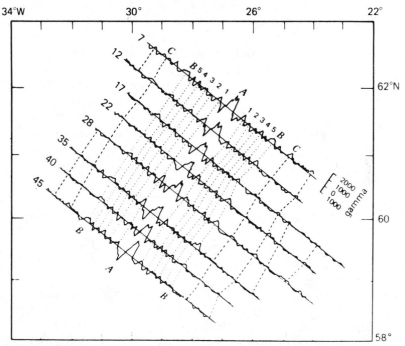

Fig.103.Magnetic anomalies over the Reykjanes Ridge, south of Iceland. Note the lineation pattern and symmetry about the central ridge axis. The central anomaly (A) has an amplitude of about 2000 γ. (After Heirtzler et al., 1966.)

Within the last decade, magnetic surveys have provided a wealth of information on the structure and the magnetic character of the oceanic crust. As we shall see later, oceanic magnetic anomalies have been a key to the understanding of the mechanism and history of sea-floor spreading (p.357).

MAGNETIC SURVEYING AND INTERPRETATION
TECHNIQUES

Magnetometer surveys are being carried out on an increasing scale for geological mapping and mineral exploration. In recent years the activity in magnetic surveying, both over continents and oceans, has become so intense that it is nearly impossible to indicate the state of global coverage. Much of the progress in this field has been possible through advances in magnetic instrumentation and data analysis. For a fairly comprehensive review of these developments, the reader is referred to the two special issues of *Geophysics* on magnetic methods edited by Affleck (1964, 1965), and to the special issue of *Bolletino di Geofysica* edited by Morelli (1970) on basement mapping by magnetics. We shall discuss here some of the important methods and techniques currently in use for measuring and interpreting the magnetic field data.

Magnetic measurements over land and sea

Almost all large-scale magnetic surveys over land are presently made by airborne magnetometers. For airborne measurements, the magnetometers in use are either the "flux-gate" type or the "nuclear-precession" type.

In flux-gate instruments, the main elements are two identical cores of a high magnetic susceptibility material (such as μ metal), which are placed parallel to each other, and wound with primary coils in series, but in an opposite sense (Fig. 104). A secondary winding surrounds the two primary windings. The primaries are connected to an a.c. source and the cores are brought to saturation in opposite directions, twice per cycle. The voltage induced in the secondary will be proportional to the net rate of the change of magnetic flux due to the two cores, this being normally zero, since the cores are changing their magnetization in opposite senses. However, the existence of an external field along the axis of the cores will disturb the flux balance by reinforcing the exciting field in one core

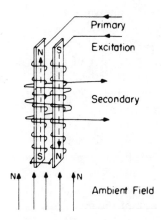

Fig.104.Principle of flux-gate magnetometer. The voltage induced in the secondary coil is proportional to the strength of the ambient field. (After Garland, 1971, fig. 16.1.)

and opposing it in the other. For a small ambient field, such as due to the earth, the secondary voltage is proportional to the strength of the field, and the device is therefore capable of measuring the field component along the axis of the cores. Instruments of this kind can be used to record any component of the geomagnetic field, depending on the alignment of the sensitive element. In airborne surveys, normally the magnitude of the total field, F, is measured, since the alignment precision in this case is less critical. The accuracy in the measurement of F is possible to approximately 1 γ which is about one part in 50,000.

Recently several portable designs for use in ground surveys have also become available on the market. In contrast to the conventional tripod-mounted field magnetometers (e.g. Schmidt-type deflection variometers or the compensation-type variometers made by Askania, Berlin), the flux-gate ground instruments are very light, need no precise levelling, and are comparatively quick in operation. The relative accuracy in the measurement of the fields by these instruments is of the order of a few gammas. One such instrument is shown in Fig.105.

In the nuclear-precession-type instruments, use is made of the well-known phenomenon of the precession of nuclear particles around a magnetic field, the frequency of precession being proportional to the strength of the ambient field. By far the most commonly used meter in airborne or shipborne surveys is the "proton" magnetometer. Here the active element is water (or some other liquid containing a large number of hydrogen nuclei) in a small bottle surrounded by a suitable coil. Normally, the spin moments of the protons are randomly oriented. If a strong magnetic field ("polarizing field") is applied by sending a direct current in the coil, the moments due to the protons become aligned in the direction of the field. When the polarizing field is suddenly removed, the spinning proton moments precess for a short time around the direction of the earth's ambient field and induce a small voltage in the coil. The frequency, f, of this voltage, which can be measured with a great precision (one part in 100,000), is a measure of the earth's total field, F. The working relation between f and F is:

Fig.105.Exterior view of a digital flux-gate magnetometer.

$$F \text{ (in } \gamma \text{)} = 23 \cdot 4868 f \qquad (4.21)$$

where f is the measured frequency in Hz and $23 \cdot 4868$ is a constant related to the gyromagnetic ratio of the proton. An important advantage of this instrument is that orientation is not critical; the only requirement is that the polarizing field should make a sufficiently great angle with the direction of the earth's total field. In contrast to the flux-gate magnetometer, which can measure the field continuously, the proton magnetometer gives a series of discrete measurements at intervals* of a few seconds because of the polarizing and relaxing time taken by protons. Recording proton magnetometers are now built by several manufacturers for station, ship, and aircraft use. Some portable designs for ground surveying are also available, e.g. Elsec (Oxford), Askania (Berlin), and Geometrics (Palo Alto, California).

Although measurements can be made as easily with an airborne magnetometer as with a ship-towed magnetometer, at present most measurements at sea are made with ship-towed proton magnetometers. The use of ship enables the magnetic measurements to be correctly related to the submarine topography, which is very necessary when specific features of the sea floor are being investigated.

Regardless the instrument used in a survey, a set of magnetometer observations generally requires some corrections of which the most important are:

(1) *Diurnal correction.* The method of applying the diurnal correction to ground survey data is analogous to that for the drift of a gravimeter. However, instead of frequent return to a base station, it is more convenient to set up an auxiliary instrument (e.g. a self-recording flux-gate or proton magnetometer) at the base station. In airborne magnetic surveys, this correction is made by flying the survey in a pattern of parallel "flight

---

*Optical pumping magnetometers (e.g. the Rubidium vapour magnetometer used in satellites and space probes) overcome this difficulty and can record the total field continuously with a precision of 1 part in a million.

lines" with a pair of "tie lines" crossing them at suitable intervals. Diurnal-variation effects are relatively less in magnitude over the oceans (except near the magnetic equator), and corrections, where necessary, can be made by checking magnetograms from the nearest observatories.

(2) *Normal-field correction.* This correction is required to take into account the normal variation of the geomagnetic field intensity with latitude and longitude. The correction is usually made by reference to national or world contour maps which give the regional values of F, H, and Z. Alternatively, tables of the I.G.R.F. (International Geomagnetic Reference Field) which are published as grid values of the total field (Fabiano and Peddie, 1969) can be used.

INTERPRETATION OF MAGNETIC DATA

The end product of a magnetometer survey is a contoured anomaly map in isogams. The next step is to translate the magnetic data into subsurface geology, and as we know from the gravity case (see section *Analysis and interpretation of gravity data*) there is no direct, short-cut approach to solving this problem. In general, the techniques employed for interpreting magnetic data are rather similar to those for gravity data. There are, however, two factors which make the interpretation of magnetic data more complicated; (1) the dipolar nature of the magnetic field, and (2) the additional unknown parameter introduced by the direction of magnetization in rocks. Nevertheless, despite these complications, magnetic surveys can give very useful geological information when applied to the right type of problems.

Qualitative interpretation

The qualitative interpretation of a magnetic anomaly map begins with a visual inspection of the shape and trend of the major anomalies. After delineation of the structural trends, a closer examination of the characteristic features of each individual anomaly is made. These features are: (a) the relative locations and amplitudes of the positive and negative parts of the anomaly, (b) the elongation and areal extent of the contours, and (c) the sharpness of the anomaly as seen by the spacing of contours. In many cases, meaningful geological information can be deciphered directly by looking at the map, without making any calculations. We shall illustrate this by an example.

Fig.106 shows a map of total field intensity anomaly, ΔT, of the Alsace–Baden region inside the Rhine Graben. Note the broad structural trends depicted by the elongation of contours, particularly the trends of the major anomalies. The negative anomaly extending in the direction Rastadt–Achern is over 40 km long and follows the axis of the graben. The main features of the positive anomalies can be explained by the existence of broad zones of high magnetization in the crystalline rocks beneath the sediments. When looking more closely at the sharp gradients around the positive anomalies, it appears that the depth to the Hercynian basement is about 2–2·5 km* below the surface. The abrupt termination of the positive anomaly contours, east of the dividing line between Bischwiller and Seltz, is suggestive of a down-faulting of the basement to a large depth towards the eastern side. No simple explanation can be given for the negative anomaly located northwest of Haguenau. Some knowledge of the magnetic properties of the crystalline rocks in this region could be of much help in attempting a quantitative interpretation.

Quantitative interpretation by models

After completing the qualitative study it is important to extract some quantitative information from the magnetic data. This process has to follow a series of steps. From the relative spreads of the maxima and minima of the anomaly we know the approximate location and horizontal extent of the causative

---

*This estimate of depth follows from a rule based on the "straight slope" parameter, see p. 219.

214

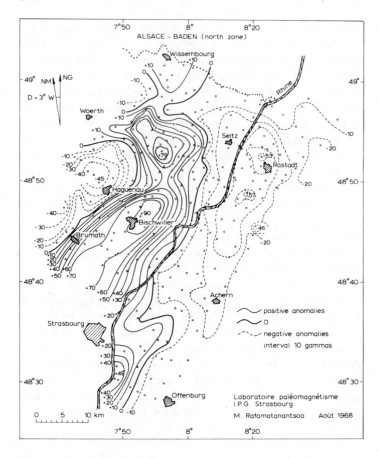

Fig.106.Map of total intensity magnetic anomalies in the Alsace-Baden region. (After Roche and Wohlenberg, 1970.)

body. Next. from the form of the anomaly, the other parameters of the body, its shape and depth, may be determined. The geometrical parameters must then be translated into structural terms in the light of known geology. Finally, from the amplitude of the anomaly, the magnetization contrast may be determined.

To develop methods for making rapid quantitative estimates of these factors, it is necessary to examine in detail the shape patterns of magnetic anomalies of the most commonly used models in magnetic interpretation. In magnetic mapping of basement structures, two models have been widely employed to approximate the shape of probable anomaly sources. One is the bottomless vertical-sided prism model (Fig.107) of rectangular cross-section. The prism model serves well to outline the volumes of rocks showing a marked magnetization contrast, and the success of this model in predicting the thickness of the non-magnetic overburden (sediments) is due directly to the fact that at sufficiently large distances potential fields are little influenced by the details of the shapes of their sources. The reader may

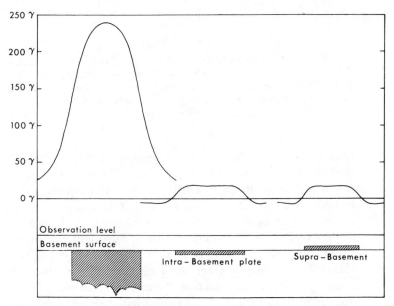

Fig.107.Magnetic anomalies caused by intra- and supra-basement bodies. The bottomless prism model (left) serves to outline magnetization contrasts extending deep below the basement. Plate models are used to represent the magnetization contrasts of limited depth extent, e.g. due to relief of basement. (After Reford and Sumner, 1964.)

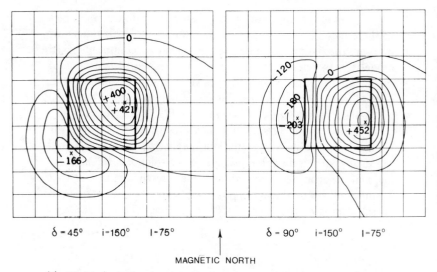

$\delta = 45°$ i=150°   I=75° $\delta = 90°$ i=150°   I=75°

MAGNETIC NORTH

Fig.108.Method of estimating the dimensions and direction of the magnetization of a prism model from a total-field anomaly (ΔT) map. The locations of ΔT_{max} and ΔT_{min} approximately mark the edges of the prism model. The direction of the magnetization is obtained by anomaly analysis as explained in the text. Total magnetic intensity model $= 3 \times 3 \times 0.25$, grid interval = depth of burial. (After Zietz and Andreasen, 1967.)

refer to the G.S.A. Memoir by Vacquier et al. (1951) for the theoretical principles involved and some examples of the application of this model to basement depth investigations. The method essentially involves comparison of the observed anomaly pattern with the computed anomaly sets in the album of the Memoir for estimating the size and depth of the model source. The Memoir also gives a method of calculating the susceptibility contrast from the amplitude of the observed anomaly on the assumption that the total magnetization of the body is in the direction of the earth's inducing field.

An alternative method suitable for block-type bodies with an arbitrary direction of magnetization vector J ($=J_i + J_r$) has been suggested by Zietz and Andreasen (1967) which is more convenient for making semi-quantitative interpretations. In this method use is made of the relative amplitudes and locations of the positive and negative parts of the anomaly to estimate the direction of the magnetization and the dimensions of the vertical prism model (Fig. 108). The ratio A_r ($= \Delta T_{max}/\Delta T_{min}$) gives a rough measure of the inclination, i, of the magnetization vector. A useful practical guide for making semi-quantitative estimates for i is: $i \sim 0°$ for $A_r \sim 1$; $i \sim (30°-50°)$ for $A_r \sim (2-5)$; $i \sim (60°-75°)$ for $A_r \sim (6-20)$. Further, if $i \gtrless 60°$ the azimuth δ of the magnetization vector is determined by the angle which the line joining the ΔT_{max} and ΔT_{min} centres makes with the magnetic north. Furthermore, the locations of the maxima and minima approximately define the edges of the prism model (see Fig.108). As to the determination of the depth to the top of the model, there are several empirical "rules of thumb" which we shall discuss later.

The other model commonly used is the two-dimensional rectangular strip (or the so-called long tabular model) which serves to approximate many geological features of extensive strike length, e.g. thick dykes and broad magnetic zones cutting across the host rocks. The formulae for the magnetic anomaly due to an infinitely long strip can be obtained in terms of its magnetization contrast, width, and depth extent. Referring to Fig.109, it is easy to see that the magnetization component parallel to the strike, J_y, produces no anomaly. On summing up the respective contributions of J_z and J_x in the vertical and horizontal direction, the corresponding anomalies ΔZ and ΔX (in SI-units) are obtained as:

$$\Delta Z = \frac{2\mu_0}{4\pi} \left[J_z (\theta_1 - \theta_2) + J_x \left(\ln \frac{r_2 \, r_3}{r_1 \, r_4} \right) \right] \quad (4.22)$$

$$\Delta X = \frac{2\mu_0}{4\pi} \left[J_x (\phi_1 - \phi_2) + J_z \left(\ln \frac{r_2 \, r_3}{r_1 \, r_4} \right) \right] \quad (4.23)$$

These expressions can be combined to give the total field anomaly, ΔT ($\lessgtr F$), in the direction of the earth's field, F:

218

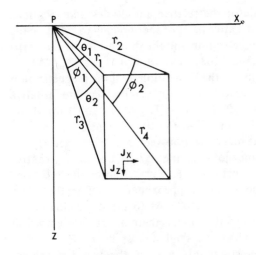

Fig.109. Quantities involved in the calculation of the magnetic anomaly over a long tabular body striking perpendicular to the page (for explanation see text). (After Garland, 1971.)

$$\Delta T \simeq \Delta X \cos I \cos \delta + \Delta Z \sin I \qquad (4.24)$$

where I is the inclination of the earth's field at the anomaly site and δ is the angle between the x-axis (perpendicular to the strike) and the magnetic north. The above equations are basic for the computation of anomalies caused by long tabular bodies with steeply dipping (nearly vertical) sides, and have been most widely used in the interpretation of the lineated anomalies over oceans (p.243). General methods for calculating magnetic anomalies due to two-dimensional and three-dimensional bodies of arbitrary shape will be found elsewhere (e.g. Talwani, 1965; Sharma, 1966, 1967).

Depth estimates from magnetic profiles

In geophysical investigations of subsurface structures, the most important parameter to be estimated is the depth to the anomalous structure. So much has been written on this subject that an attempt to review all the depth rules so far devised would justify writing a monograph on this subject itself. The crux of the problem is that a general rule for depth determination does not exist. If some simple regular shape for an anomalous body can be postulated, it is possible to devise fairly precise depth rules in terms of "half-width" or some other measure of the anomaly gradient (see Parasnis, 1972, p. 18).

For other models such as the vertical prism and long tabular prism, some empirical depth rules have been devised which, when applied with care, may furnish useful results. Of these, we shall mention only two which have been most widely employed. These are based on the "maximum-slope" and "half-slope" parameters of the anomaly profiles. Referring to Fig.110, the distance, s, is the horizontal interval over which the steepest part of the anomaly curve is substantially a straight line. This parameter s (usually called the straight slope) is related with the depth to the top of the basement block showing the magnetization contrast. The depth index cannot be defined mathematically, but empirically it is found usually to range between 0·9 and 1·3, roughly averaging to 1·1 times the depth. On isogam contour maps the straight-slope parameter is most convenient to use for the steepest part of the anomaly. The "half-slope" parameter is the distance between the points at which a straight line with half the maximum slope is tangent to the anomaly curve below and above the straight-line slope (Fig.110). The half-slope parameter, p, is related to the depth by a factor which is usually between 1·5 and 2 times the depth. For a detailed discussion of the various depth rules, the reader may refer to papers by Vacquier et al. (1951), Naudy (1970), and Åm (1972). These basic rules provide semi-quantitative estimates of depth which are useful in testing the initial geological assumptions of the model and, if necessary, in adjusting the model parameters with the aid of a computer to obtain a best fit between the observed and computed anomaly curves.

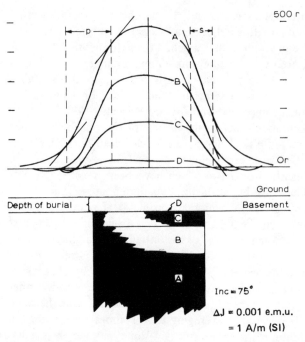

Fig.110.Amplitude and shape characteristics of a magnetic anomaly due to a block model with a variable depth to the bottom. The "straight-slope" (*s*) and the "half-slope" (*p*) parameter are used as depth indices. (Modified from Steenland, 1965.)

Computerized methods of interpretation

Recently several machine methods involving mathematical processing of anomaly data have been devised; these are becoming increasingly popular with large-survey organizations for routine analysis of magnetic data. Various computerized techniques are now available which enable a series of operations to be made automatically, for example, the contouring of anomaly maps from regularly or irregularly spaced data, the construction of residual and second derivative maps, the plotting of major anomaly profiles, estimates of size and depth parameters for a series of assumed models, the calculation of magnetization contrasts, and finally the plotting of basement-depth contours.

Undoubtedly, these machine methods are proving their utility in testing the various assumptions of the probable model sources, and in expediting the process of "indirect" interpretation by the successive adjustment and "least-squares" fitting of the model parameters. However, they do not necessarily enhance the reliability of the interpretation through these elaborate operations. Despite the various claims made about the ingenuity of automatic interpretation techniques, it must be emphasized that unless adequate geologic control or some other independent information is available, no interpretation technique, however sophisticated, can provide a unique solution.

MAGNETIC MAPPING OF GEOLOGICAL STRUCTURES

Magnetic surveys have been used to study a great variety of geological structures ranging in depth and size from deep-basement blocks to near-surface ore bodies. Until recently their most extensive use has been in oil exploration for outlining areas of maximum thickness of potentially oil-bearing sediments, and, on a smaller scale, in iron ore prospecting. However, in the last 10–15 years regional magnetic surveying, both over continents and oceans, has been made on such an enormous scale that it is difficult to have a precise estimate of the total coverage. For our purpose we shall discuss here a few examples to demonstrate the particular applicability of magnetics to problems of geological mapping.

Basement mapping under sedimentary cover

The sedimentary cover prevents the direct examination of the basement in many areas. Gravity and seismic results are to an appreciable degree influenced by the sedimentary section which, however, is virtually transparent to magnetic forces. Therefore, the magnetic method is particularly suitable to map basement structures concealed beneath a thick cover of sediments. Since sediments are practically non-magnetic, any significant anomaly observed over a sedimentary basin should have its origin at or below the basement surface. Magnetization contrasts below the basement extend usually to some tens of kilometres, and are, therefore, taken to be bottomless. Magnetic anomalies due to such deeply extending magnetic contrasts are, as a rule, substantially large, and are called "intrabasement anomalies". On the other hand, anomalies of structural origin (e.g. due to the relief of the basement) are fairly small and of limited extent (see Fig.107). These are referred to as "supra-basement" anomalies for which the interpretation model is a rectangular plate. Regional-residual methods (discussed earlier, p. 135) or second-derivative methods (Henderson, 1960) can be used for separating the two types of anomalies. In most cases the amplitude of the anomaly itself is a good index for the distinction of the two types.

Fig.111A shows the aeromagnetic anomaly map of the total field intensity in Pulasky County, Indiana. The basement is a southward extension of the Precambrian shield of Canada covered by a Palaeozoic sedimentary section. The anomaly is one of the most isolated and locally intense magnetic features in the Indiana basement complex. Comparison of this anomaly with model anomalies yields the maximum depth, the approximate areal outline, and the susceptibility contrast of the anomalous body. Fig.111B shows the block model of dimensions 4 × 6 depth units from the album of Vacquier et al. (1951) that best fits the observed anomaly. A maximum depth of 1·3 km below the surface (1·7 km below the flight level) is determined for the anomalous basement body. This agrees closely with a depth of 1·4 km obtained from seismic measurements in this area. The straight-slope method (see p.219) gives a depth estimate of only about 1·2 km. The inclination of the geomagnetic field at Pulaska is 72° which is very close to that of the model used for comparison. Both the observed and model anomaly contours show a subdued magnetic low to the north which is a typical characteristic of anomalies caused by magne-

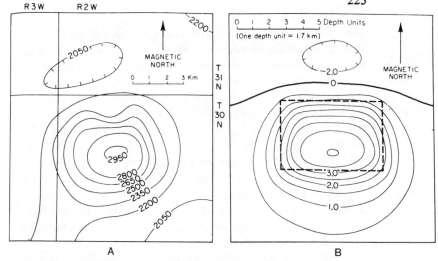

Fig.111.A. Map showing the total aeromagnetic intensity in Pulaski
County, Indiana. Contour interval is 150 γ. Inclination of geomagnetic
field is 72°. (After Rudman and Blakely, 1965.)
B.Map showing the total magnetic intensity computed for 4 × 6 depth-
unit model (dashed line). Geomagnetic field inclination is 75°. (After
Vacquier et al., 1951, p.133.)

tization along the direction of the earth's field (in the northern
hemisphere). Assuming a uniform magnetization contrast
induced by the earth's field, a susceptibility contrast of 0·003
in electromagnetic c.g.s. (or $4\pi \times 0\cdot003$ in SI) is obtained for
the 4 × 6 depth-unit model. This value should be taken as a
minimum value because it is computed on the assumption of a
block model of infinite depth extent.

For more examples of magnetic basement mapping in sedimentary areas the reader is referred to Steenland (1965) and
Nettleton (1971).

Mapping of intrusives and extrusives

The magnetic method is very effective for mapping various
types of intrusives and extrusives whose magnetic properties
are appreciably different from that of surrounding rocks. How-

ever, particularly for basic and ultrabasic rock bodies the remanent magnetization J_r is usually strong and as a rule has to be taken into consideration in the interpretation of anomalies.

On the island of Bornholm some Precambrian diabase dykes are exposed along the northern coast, but at some distance farther inland they are not visible. By magnetometer surveying it has been possible to trace the location and to estimate the depth of burial of the prominent Kjeldseå dyke up to 15 km inland. Fig.112 shows the vertical component of the anomaly (ΔZ) over the dyke and the model used in estimating its depth of burial. Since the intensity of J_r was comparable to that of J_I, the magnetization contrast, J, used in the model calculations was obtained by a vectorial addition of J_r and J_I. Schønemann (1972) has made detailed studies of the magnetic properties and the palaeomagnetism of this dyke. From a detailed magnetic ground survey on East Bornholm, Platou (1970) has been able to demarcate the zones of contact between granites and gneisses. Fig.113 illustrates the granite–gneiss contact inferred from the magnetic anomaly map.

Bodies of basic igneous rocks seated in an environment of sediments are characterized by an assemblage of anomalies that can be easily separated. This can be seen in Fig.114 which represents a section of a regional anomaly map in the Eifel Mountains of Germany. The extensive circular anomaly evidently is produced by a large basic pluton. The numerous anomalies of high amplitude and small extent belong to outcropping basalt pipes.

A further example of magnetic mapping of mafic rock bodies can be seen in Fig.115 which shows the aeromagnetic and generalized geology map of the Wild Horse Ridge area in Montana. Note the major structural trend of the volcanic ridge which is delineated on the magnetic map. The ridge is dominated by a large positive anomaly on the north side and an equally large negative anomaly on the south side, indicating a prominent remanent magnetization direction substantially different than that of the earth's field. The remanent magnetiza-

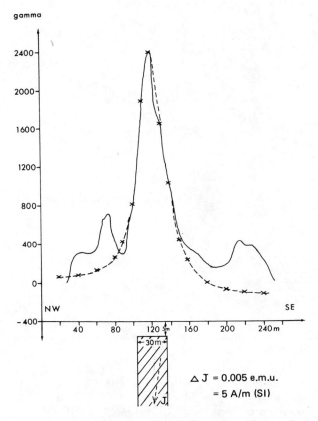

Fig.112. Observed Z component of the magnetic anomaly (continuous curve) over the Kjeldseå dike, Bornholm, and the computed interpretation model (dashed curve with crosses). (After Schφnemann, 1972.)

tion is so dominant that the contribution of induced magnetization to the anomaly can be neglected. The average field intensity (relative to an arbitrary base) in the vicinity of the ridge
area is nearly 4000 $\dot{\gamma}$ and, therefore, the amplitude ratio, A_r
($= \Delta T_{max}/\Delta T_{min}$), is about 1. Referring to the quantitative
significance of A_r (see p.217),   it is seen that this ratio corresponds to an inclination of the magnetization vector i of zero

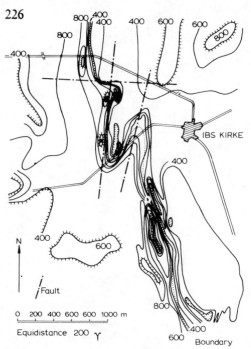

Fig.113. Part of the magnetic anomaly (vertical intensity) map of East Bornholm. The Svaneke granite–gneiss boundary is marked by a minimum trench of about 400 γ. (After Platou, 1970.)

degrees. The azimuth, δ, of the magnetization vector as given by the angle of the line joining $\Delta T_{\max}$ and $\Delta T_{\min}$ on the aeromagnetic map and the magnetic north is 320°. These inferred values of $i = 0°$ and $\delta = 320°$ from the anomaly analysis compare favourably with those obtained from magnetic measurements on the collected samples. This example illustrates the potential use of anomaly analysis of volcanic bodies for palaeomagnetic applications, especially in those cases where $J_r \gg J_i$ The strong remanence associated with volcanoes and seamounts has been used in a number of cases to determine the fossil field direction (Vacquier and Uyeda, 1967; Sharma, 1969) of the geomagnetic field prevailing at the time when the volcanoes were formed.

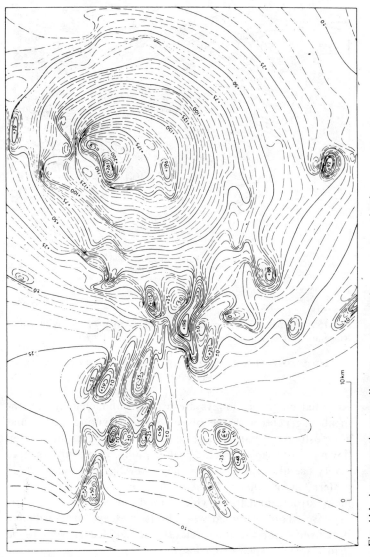

Fig. 114. Aeromagnetic anomalies (in γ) measured over an area in the
Eifel Mountains, W. Germany. (After Hahn, 1971.)

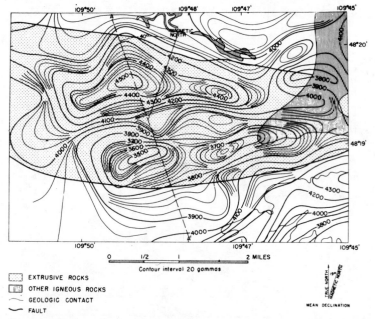

Fig.115. Aeromagnetic and generalized geology map of the Wild Horse Ridge area, Montana. (After Zietz and Andreasen, 1967.)

AIRBORNE SURVEYS FOR REGIONAL GEOLOGY AND TECTONICS

Regional magnetic surveys are usually conducted by governmental agencies with the objective of providing quick and composite information over large areas, and are now often flown prior to regional geological mapping. These surveys serve a very useful purpose in that they show the regional geologic pattern, the magnetic character of different rock groups, and major structural features which would not be noted if the survey covered only a limited area. This is an exceptionally valuable background for the interpretation of detailed surveys to be made later for special purposes.

An example is provided by the regional survey flown over Scandinavia. Fig.116 shows a part of the regional aeromagnetic

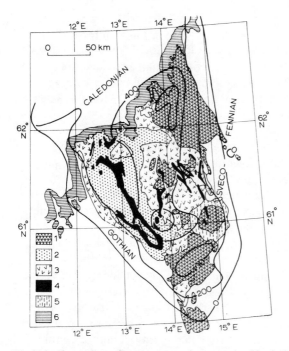

Fig.116. Kopparberg County magnetic anomaly. Vertical field magnetic contours at 200-γ intervals from the profiles shown. Flight altitude was 3000 m. Geology: 1=sub-Jotnian Dala granites; 2=Jotnian sandstones; 3=sub-Jotnian volcanics; 4=Jotnian diabases; 5=Caledonian front; 6=Ordovician and Silurian sediments. (After Riddihough, 1972.)

map of Fennoscandia over the Kopparberg County in west-central Sweden. The anomaly feature shown here is a southern part of a large magnetic high extending beyond the Caledonian front. The geological feature which forms the source of this anomaly is probably continuous northwards at a greater depth beneath the Caledonian rocks (as shown by a notable drop in the amplitude of the anomaly across the Caledonian front). The geological correlation between the magnetic high and the outcrop of the "Jotnian" complex can be seen clearly in the figure. Both tectonically and magnetically the whole complex can be visualized as a down-faulted basin structure of rocks

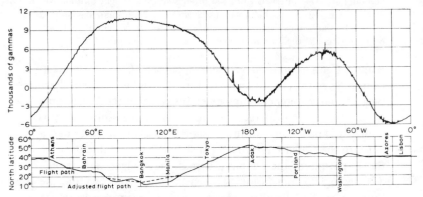

230

of volcanic origin with numerous granitic intrusions. From magnetic model comparisons Riddihough (1972) has estimated the probable depth extent of the Jotnian and sub-Jotnian rocks to be about 20 km with a magnetization contrast of approximately 0·003 e.m.u./cm^3 (or 3 A/m in SI) for the extrusive volcanic rocks.

A further example of a large-scale airborne survey is provided by around-the-world magnetic profile made on a very long flight line (Alldredge et al., 1963). Fig.117 shows the position of the flight path and the observed anomaly, ΔT, of the total field intensity. The magnetic profile depicts the major regional variations of wavelengths of several thousands of kilometres whose sources are certainly very deep seated (probably at the core–mantle boundary, see p.181). On these broad anomalies are superposed local perturbations of widths which are at most a few hundred kilometres. They evidently are of shallow origin, at depths of some tens of kilometres in the earth's crust. The conspicuous absence of anomaly features of intermediate depth clearly shows that there are no magnetic field sources in the mantle. From the sharpness and amplitude of the local perturbations (anomaly kicks) the following general

Fig.117. Around-the-world magnetic profile showing both the positions and non-dipole fields (geomagnetic anomaly). The flight altitude was 9000 ft ($\approx$2700 m). (After Alldredge et al., 1963.)

suggestions were inferred (Alldredge et al., 1963):

(1) The profile over the entire Mediterranean Sea is nearly void of local anomalies except over Sicily and the toe of Italy. This probably indicates a great depth to basement rocks.

(2) Deep magnetic sources with high magnetization contrasts are indicated under India, whereas across southeastern Asia the sources are relatively shallow and of low magnetic contrasts.

(3) Across the entire Pacific, from Tokyo to Portland via Adak, the sources are shallow. The structure of magnetic basement from the Aleutian trench to Portland is broken up into small, quite regular pieces.

(4) The major magnetic sources across the United States tend to be very deep, with numerous shallower intrusions.

(5) The magnetic sources are very shallow over a region from about 40°W to Lisbon. These shallow sources are centred near the Azores and are associated with the Mid-Atlantic Ridge.

PALAEOMAGNETISM AND ITS APPLICATIONS

Palaeomagnetism or "fossil magnetism" is primariy concerned with the study of the natural remanent magnetization (NRM) of rocks in order to provide reliable information about the earth's magnetic field in geological times. The great advances in this field over the past 20 years have been largely stimulated by Professors P. M. S. Blackett and S. K. Runcorn in England, Professor E. Thellier in France, and Professor T. Nagata in Japan. Results from this comparatively recent field of geophysics have already greatly extended our knowledge about the history of the geomagnetic field (including its polarity reversals). However, to-date the most exciting applications of palaeomagnetism have been in providing quantitative evidence on such fascinating hypotheses as continental drift, sea-floor spreading, and plate tectonics. In recent years palaeomagnetism has been increasingly used as a structural and correlation tool in the study of local geological problems.

An excellent general account of palaeomagnetism is provided in Irving's (1964) book; a good summary appears in Tarling (1971). Techniques are described in a volume edited by Collinson et al. (1967). Reviews incorporating more recent data and developments can be found in Creer (1970) and in McElhinny (1973). Within the scope of this book, it is possible to outline only the basic principles and most important applications of palaeomagnetism.

Methods of palaeomagnetism

The fundamental basis of palaeomagnetism is the fact that the stable NRM of a tectonically undisturbed rock unit provides a faithful record of the ancient geomagnetic field (F_{anc}) prevailing at the time when the rock was formed. The basic postulates are expressed as $J_r//F_{\text{ane}}$ and $J_r \propto F_{\text{anc}}$; for some significant applications on a global scale, the additional assumption is made that F_{anc} conforms to a geocentric axial dipole. The latter implies that the geomagnetic dipole axis, on average, corresponds with the earth's geographical axis. The basic postulates appear to be substantially supported by ample palaeomagnetic data obtained from different continental and oceanic areas.

We have mentioned earlier (see section *Magnetization of rocks*, p.200) that mostly TRM and CRM, being comparatively more stable than other kinds of NRM, have been used as basic data for palaeomagnetism. As to the methods in palaeomagnetism, the first requirement is the collection of a set of samples, oriented in space, from the rock unit to be studied. The NRM of a rock outcrop, especially in areas of basic extrusives (and/or intrusives), may seriously affect the compass readings, and therefore a sun compass (Creer and Sanver, 1967) may often have to be used instead. If it is obvious that the rock formation has undergone a deformation, as in the case of a dipping layer, the original horizontal plane as indicated by bedding, is also marked. For a layered sequence (e.g. lava flows), it is usual to take samples from a considerable vertical section corresponding to some thousands of years in time, so that, when the measured NRM of samples is averaged, the effect of the secular change in F_{anc} is minimized.

From each sample thus collected, a number of specimens are cut or drilled in the form of small cubes or cylinders. For measuring the NRM of specimens, two types of instruments are in current use, the astatic magnetometer (illustrated in Fig.118) and the spinner magnetometer (Fig.119). Their sensitivities are comparable ($\sim 10^{-4}$ A/m in SI*), and both types are now available as commercial units. A specimen is placed in various orientations in either instrument to determine its NRM vector. The measured NRM vectors of rock speci-

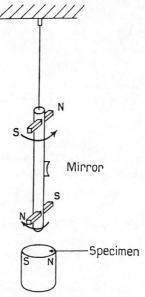

Fig.118. Principle of the astatic magnetometer. The deflection of the suspended magnet system is a function of the distance, direction and intensity of the magnetization of the specimen.

*The SI-unit of magnetization is A m, which equals 10^{-3} e.m.u./cm^3 (see Appendix 1).

Fig.119.The rotating head of a spinner magnetometer. The rotation of a rock specimen within the pick-up coil induces a voltage in the coil, the phase of which depends on the direction of the remanent magnetization and the amplitude on the intensity.

mens yield a number of directions in space which can be conveniently represented by a stereographic projection. The grouping of points representing the direction vectors on the stereogram then expresses the degree of consistency in the determinations.

Several geological tests have been applied to determine the stability of the NRM, of which the "fold test" and the "baked contact" test are most fundamental. The principle of the fold test (Graham, 1949) is illustrated in Fig.120. If the directions of NRM in a folded bed, sampled at different locations, differ from each other, but are brought into agreement after applying the tilt corrections, the NRM predates the tectonic disturbance and has remained stable since that time. The baked contact test makes use of the fact that when an igneous melt intrudes into a host-rock, the latter is heated, and upon cooling acquires an NRM (presumably TRM) in the same magnetic field in which

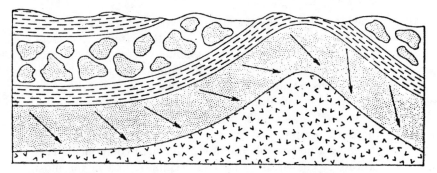

Fig.120.Graham's (1949) fold test. If the directions of remanent magneti-zation (arrows) become uniform after applying tilt corrections, the remanent magnetization predates the tectonic disturbance.

the intrusive rock on cooling becomes magnetized (Fig.121). Since the igneous (or baking) rock and the baked country rock are generally different materials, agreement between their NRM directions and disagreement with the NRM direction of the unbaked rock provides good evidence for the stability of magnetization of the intrusive rock. This situation also applies to the baked rock underlying an extruded lava flow. As a matter of fact this test has provided very convincing evidence for the reality of the polarity inversions of the geo-magnetic field, as demonstrated by 87 agreements between baked and igneous senses out of 90 reported baked contacts (R. L. Wilson, 1966).

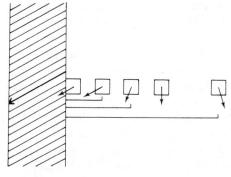

Fig.121.Schematic illustration of the zones of magnetization after cooling of the igneous rock and the adjacent baked rock (After R. L. Wilson, 1966.)

Before interpreting the measured NRM of a rock in terms of the direction of the ancient field, the viscous remanent magnetization (VRM) or other secondary magnetizations that are frequently superimposed on the primary remanence have to be carefully removed. These secondary magnetizations are usually (but not always) "soft" compared to the primary TRM or CRM, and in favourable cases it is possible to destroy the. soft components by a partial demagnetization while retaining a measurable fraction of the hard component. This process is known as "magnetic cleaning", and in the most common method the rock specimens are subjected to an alternating field which smoothly decreases to zero from a selected peak intensity depending on the coercive force, H_c, of the components to be destroyed. Alternatively, rocks can also be "cleaned" thermally by stepwise heating and cooling them in a field-free space. The thermal method is normally less convenient than the alternating-field method, but can be more useful when rocks have undergone complex thermal histories and acquired secondary TRM or PTRM.

A consistent set of directions after the cleaning operation is taken to be the "fossil" remanence, indicative of the ancient field direction. Strictly speaking, this interpretation should be accepted with great caution. Various other checks to detect any post-formation changes in rocks that can be caused by a subsequent geological (or tectonic) event, or by physico-chemical alterations in the magnetic minerals, have to be made in order to rule out the possibility of any secondary hard component (TRM or CRM) acquired during a remagnetization process (Storetvedt, 1968).

The mean direction of NRM from a consistent group can then be used to infer the equivalent pole position, on the assumption of an axial dipole field. The necessary formulae for this transformation can be found in any palaeomagnetic text (e.g. Irving, 1964). Furthermore, since the magnetization directions on the original stereogram define an area, rather than a point, the ancient pole (or palaeomagnetic pole) is obtained as an area on the globe. Various statistical methods to determine the size and shape of this area in terms of confidence limits have also been discussed by Irving. Palaeomagnetic pole positions determined from rocks of Pliocene and Pleistocene age are shown in Fig.122. The cluster of poles about the geographic pole (leaving the present geomagnetic pole very much apart) clearly indicates that the geomagnetic

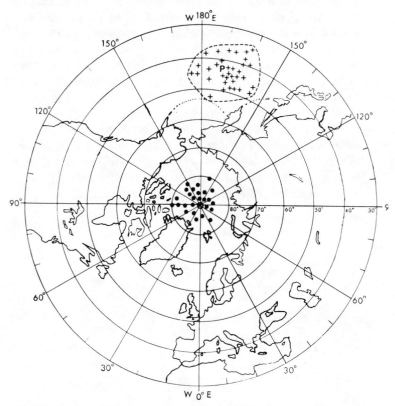

Fig.122.Stereographic projection of the northern hemisphere. Black dots clustered around the North Pole indicate the positions of the Pleistocene and Pliocene poles determined palaeomagnetically. *M* indicates the present position of the geomagnetic pole. The significance of the Permian poles (clustered around *P*) is discussed on p.247. (Redrawn from Holmes, 1965, fig.865.)

field has been, on average, an axial geocentric dipole field at least during the recent past.

Reversals of the geomagnetic field

One of the most significant discoveries in palaeomagnetism is that about 50 % of the rocks studied so far are found to be magnetized in a direction approximately opposite to that of the earth's present field. This is attributed to the fact that the earth's field has reversed its polarity many times in the geological past. The fifty-fifty occurrence of normal and reversely magnetized rocks indicates that the earth's field may have either polarity with equal probability; if a self-reversing magnetization (p. 199) has occurred in almost 50 % of the rocks, it would be a rather remarkable coincidence.

The reality of geomagnetic field reversals has been confirmed by independent evidence from "catching the field in the act of reversing" in lavas of Iceland (Brynjolfsson, 1957) and in deep-sea cores (Fig.123) from the North Pacific (Opdyke, 1968). Another convincing piece of evidence is from the

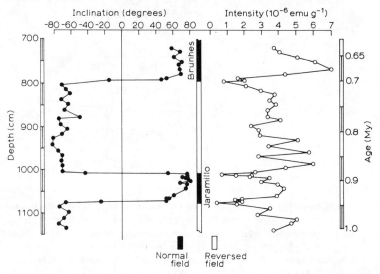

Fig.123.Variation of the intensity of magnetization and the inclination with depth as recorded in a deep-sea sedimentary core.(After Opdyke, 1968.)

"baked contact" polarity test (R. L. Wilson, 1966) which has previously been mentioned. The polarity inversions are observed to be abundant in all geological periods, with the exception of the Permian, which has been found to be a long period of predominantly reversed polarity.

Detailed studies of the polarity reversals have been made for the past $4\frac{1}{2}$ m.y. (Fig.124). These have shown the occurrence of world-wide major epochs lasting about 1 m.y. with brief polarity events or excursions within these epochs lasting only 10,000–100,000 years. The most recent reversal (Laschamp event) occurred probably only about 20,000 years ago; this event might be contemporary to the Gothenburg event discovered recently in Swedish varves (Mörner et al., 1971). Polarity inversions have been widely used as a tool in the stratigraphical correlation of deep-sea sedimentary cores (Fig.125) and of lava flows. It should now be easy to correlate contemporaneous sea-floor sediments over the world.

While undergoing a polarity transition, the earth's field intensity, F, decreases to about 15% (or less) of the normal field intensity (Watkins, 1969). Estimates of the duration of the low field intensity vary from 1000 to 10,000 years (see Fig.123) and during this period the low magnetic field of the earth could possibly cause significant biological and climatological changes. In this context, some interesting studies have been made, and a direct causal relationship between geomangnetic reversals and biological extinctions has been proposed (Crain, 1971). Some correlation has also been observed between the geomagnetic field intensity variations and the climatic changes. For instance, higher magnetic field intensities seem to correlate with the colder climates prevailing during the ice ages (Wollin et al., 1971). The statistics of these correlations are at present very poor, and no definite conclusions can be drawn.

Concept of sea-floor spreading

An application of geomagnetic field reversals with very far-reaching implications in the field of earth sciences has

240

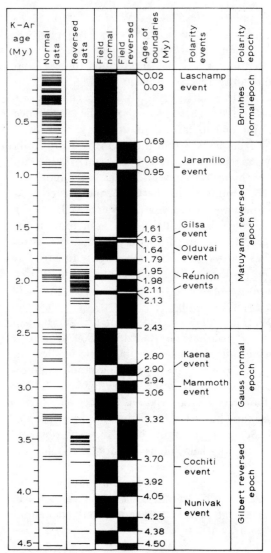

Fig.124.Chronology of the geomagnetic field reversals from K–Ar age determination on lava flows. (After Cox, 1969, with some additional data from McElhinny, 1973.)

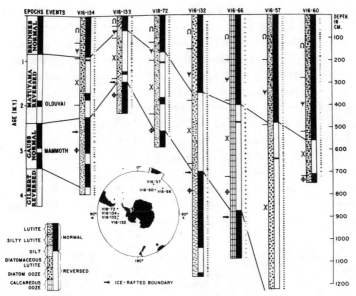

Fig.125.Correlation of the magnetic stratigraphy in seven cores from the Antarctic; locations are shown in the inset. Boundaries between the fossil zones are shown to the left of each section. (After Opdyke, 1968.)

emerged from the magnetic surveys of the oceans. As mentioned previously (p. 205), these surveys have revealed linear magnetic anomalies parallel to the ridges in all oceans. These anomalies have been interpreted in terms of alternating linear bands of normal and reversely magnetized basalt in layer 2 of the oceanic crust, and they are now recognized as direct evidence of sea-floor spreading from mid-ocean ridges.

The hypothesis, now known as "sea-floor spreading", developed from the suggestions of Hess (1960, 1962) and Dietz (1961) that new crust is progressively formed by magmatic processes in the crestal zone of ocean ridges.* The sea-floor spreads laterally in both directions to accommodate the newly

*Hess' and Dietz's hypotheses regarding the formation of new oceanic crust by magmatic processes are discussed in Chapter 8, section *Ocean-floor tectonics.*

formed crust, implying the drifting of continents (e.g. North America and Europe) as a consequence of this spreading. Vine and Matthews (1963) unified the idea of sea-floor spreading with periodic reversals of the geomagnetic field. According to their model (Fig.126), oceanic layer 2 has acted as a magnetic tape, recording the polarity of the geomagnetic field as newly formed basaltic material is driven up by convection and rises along the crest of a ridge and spreads outwards. This crustal "magnetic tape" can be replayed by observing the magnetic anomalies along the profiles crossing the ocean ridges. Because of the high Q values for oceanic basalts ($J_r \gg J_i$) the alternating strips of observed magnetic anomalies can be directly interpreted to delineate the zones of normal and reverse remanent magnetization in oceanic layer 2 (see Fig.127).

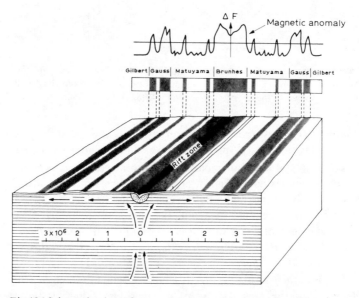

Fig.126. Schematic view of ocean-floor spreading as evidenced by positive and negative anomaly sequences which are caused by normally and reversely magnetized sections of the oceanic crust. Normal polarity zones are shaded. (After Allan, 1969.)

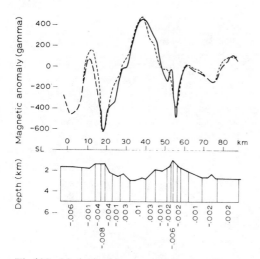

Fig.127. Method of interpretation of lineated magnetic anomalies found over mid-ocean ridges. The observed anomaly (solid line) can be simulated by a model consisting of normally and reversely magnetized blocks of oceanic crust. The figures beneath the model show the lateral change in direction and strength of magnetization. (After Loncarevic et al., 1966.)

Detailed comparison between the linear magnetic anomalies observed over the oceans and the geomagnetic polarity scale, now well established for the past 5 m.y., indicates the rate of spreading along several ridges to be from about 1 to 5 cm/year per ridge flank (Fig.128). The characteristic linear magnetic features found in the world's oceans are shown in Fig.129, which leads us to the startling observation that perhaps 50% of the present deep-sea floor, i.e. one-third of the surface area of the earth, has been created during the last 70 m.y. representing the most recent 1·5% of the geological time scale. The age of the older anomalies (>5 m.y.) shown in Fig.129 has been estimated by assuming the same spreading rates as those determined for the last 5 m.y., and these age estimates can be checked against the age of the oldest sediments overlying the igneous ocean floor using the microfossil content of drill cores. These sedimentary dates have confirmed,

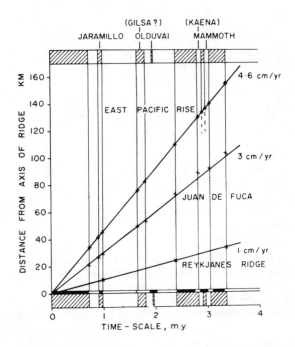

Fig.128.Inferred normal-reverse boundaries in the oceanic crust plotted against the suggested reversal time scale (shaded intervals indicate the periods of normal polarity). The thin black-and-white time scale is that suggested by Cox (1969). (After Vine, 1968.)

within 5–10%, the ages estimated from a uniform spreading rate of 2 cm/year in the South Atlantic (Fig.130), so that the anomaly pattern of the ocean floor can be used to extrapolate the time scale of the polarity changes over the last 80 m.y. or so (see Fig.181).

An independent line of evidence supporting the spreading hypothesis has been advanced by Wilson (1973), who claims that the volcanic islands of the Atlantic were originally formed at the mid-oceanic ridge and that they are now being carried away by the moving sea-floor. The age of islands should

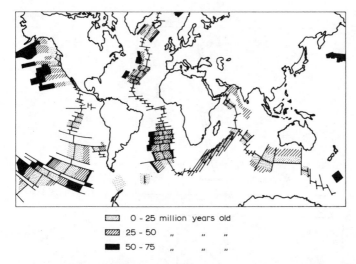

☐☐☐ 0 - 25 million years old
▨▨▨ 25 - 50   „   „   „
■■■ 50 - 75   „   „   „

Fig.129. Simplified isochron map of the ocean floor according to the magnetic anomaly pattern. Thin lines represent fracture zones. (Based on Heirtzler et al., 1968; redrawn from Tarling, 1971.)

therefore increase with increasing distance from the ridge. The ages of Atlantic islands, which have been determined so far, support the general prediction of Wilson, but cannot be reconciled with the assumption of a uniform spreading rate. Also in the northwestern Pacific, palaeomagnetic studies of a number of seamounts (Vacquier and Uyeda, 1967) suggest that these seamounts have migrated a long distance from the south. The migration of seamounts without the migration of the sea-floor itself is inconceivable.

Until the actual movement of the sea-floor can be directly detected, there will be some uncertainty about the Vine–Matthews model. The uncertainty centres around: (1) the petrological and magnetic characters of oceanic layers 2 and 3, (2) the subjectivity in the choice of models for interpreting the marine magnetic anomalies (Watkins, 1968), and (3) the driving processes and mechanisms of lithosphere consumption. Beloussov(1970) has elaborated on some of the major diffi-

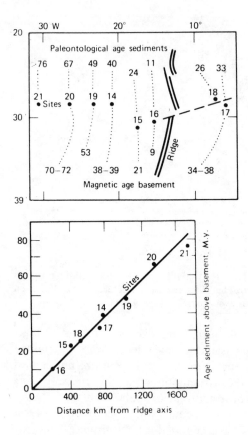

Fig.130.Comparison of palaeontological and magnetic anomaly ages from deep-sea drilling in the South Atlantic. The palaeontological age of the sediment immediately above the basalt is plotted against the distance from the ridge axis. The index number of the points refers to the drilling sites. (After Maxwell et al., 1970.)

culties relating to the spreading hypothesis and he thinks that the alternative concept of "basification" (oceanization) of the continental crust is more plausible. The author is of the opinion that the magnetic evidence, although not equally consistent for all the oceanic areas, is generally very impressive.

The magnetic evidence for sea-floor spreading, especially in light of the recent results of the JOIDES (Joint Oceanographic Institutes Deep Earth Sampling) and DSDP (Deep-sea Drilling Project) has become so compelling that it is difficult to imagine how the evidence could be interpreted otherwise. As we shall see later, this elegant idea has added a new dimension to the evaluation of the mechanism and history of continental drift and thereby paved the way for the "revolutionary" unifying concept of plate tectonics (see Chapter 8, p.366).

Polar wandering and continental drift

Palaeomagnetism has provided an independent line of evidence in the resolution of the most debated issue of the century, namely continental drift. In dealing with this problem, we are concerned with a time scale of many millions of years in order to correlate palaeomagnetic data with the history of continents. On this time scale the polarity of the geomagnetic field is of no significance, but the basic question as to whether it had an axial dipole character in the geological past assumes a key importance.

There are two lines of evidence suggesting that the earth's field has corresponded to that of an axial dipole in the remote past: palaeomagnetic and palaeoclimatic. We have shown previously (Fig.122) that the results of palaeomagnetic data on rocks of Pleistocene and Pliocene age convincingly demonstrate the geocentric axial dipole character of the earth's field. In data from older rocks, tectonic movements become increasingly significant, and for periods older than the late Tertiary the grouping of the poles becomes poorer on an intercontinental scale. However, on a continental or subcontinental scale the evidence can be extended to older periods. Fairly tight grouping of the Cretaceous poles from widely separated areas in North America (Strangway, 1970), and of the Triassic and Permian poles from the European and Siberian U.S.S.R. (Khramov and Sholpo, 1967) give ample support to the hypothesis that the geomagnetic field had been predominantly dipolar for at least 300 m.y. The correspondence between the average dipole axis and the geographical axis is, however, difficult to establish by palaeomagnetic data alone, and must be tested using some other indicators of ancient latitudes. Blackett (1961) has compiled palaeoclimatological data showing that the geographic poles in the past were very far from the present earth's rotation axis. He furthermore found that the palaeogeographic latitudes estimated from the palaeoclimatic data are consistent with those determined from palaeomagnetic data (see Fig.132). This correlation between ancient climates and palaeomagnetic latitudes together with the above mentioned consistency of the Permian and Triassic poles, appears to support the geocentric dipole hypothesis at least from the late Palaeozoic to the present. The validity of this hypothesis has, however, been questioned by Briden (1970a); his analysis of the Permo-Triassic data suggests the existence of substantial non-dipole field components during that period.

There are two principal ways of presenting palaeomagnetic data for a given region over a number of geological epochs. In one approach, which is the simpler and more useful, the poles determined for the region from epoch to epoch are plotted on the present globe, or on a stereographic projection of it. A curve through these poles is referred to as a "polar wander curve" (for that particular region), a dated path which the pole may have followed. This apparent polar movement could represent polar wandering or continental drift and the data from a single continental region alone cannot distinguish between the two. The rate of apparent polar wandering appears to average about $0.3°/$ m.y. (~ 3 cm/year), but it may be quite irregular. Note the striking agreement in order of magnitude between the polar wander rate and the rates deduced for sea-floor spreading.

Comparison of the pole paths for different continents reveals a major disagreement, as shown in Fig.131. The fact that the polar wander curves for different continental blocks do not agree is essential palaeomagnetic evidence for continental

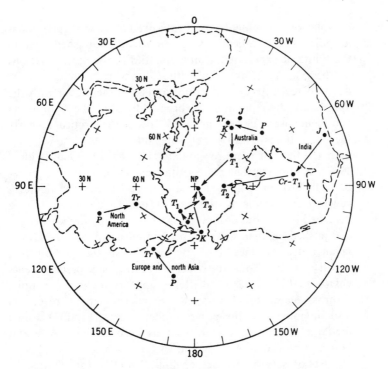

Fig.131.Comparison of the pole paths for four continental blocks since the Permian period. *NP* is the present North Pole. The palaeomagnetic poles are means for the geological periods: Permian (*P*), Triassic (*Tr*), Jurassic (*J*), Cretaceous (*K*), Lower Tertiary (T_1) and Upper Tertiary (T_2). (After Irving, 1964.)

drift. The curves can only be reconciled by a relative movement of the continental blocks over thousands of kilometers. This may not be a unique solution to the geometrical problem, but if the axial dipole hypothesis is accepted, the difference between the polar wander curves can only be interpreted as continental drift. The most important results in this context are an apparent westward drift of North America by perhaps 4000 km since the Carboniferous, a rapid northward movement of India over 50° of latitude since the Cretaceous, and a complicated movement of Australia around the south geographic pole. These indications appear to be established beyond the limits of experimental uncertainties in the determinations of the pole positions, and although the palaeomagnetically deduced reconstructions do not agree in exact detail with the Wegnerian drift picture, the broad idea that the Gondwana continents were closely grouped around Antarctica in early Mesozoic times is fairly well substantiated.

In the light of more recent data (J.T. Wilson, 1966; Briden, 1967), it appears that continental masses may have been repeatedly broken up and rejoined in several ways. A particularly suggestive example is the Indian peninsula, which having detached itself from Gondwanaland moved rapidly northward and pushed up against Asia; the continued uplift of the Himalayan range (Holmes, 1965, p.596) suggests that the pushing is still going on. Thus, the concept of primeval supercontinents envisaged by the early "drifters" appears to be too simple. A gross oversimplification of the palaeomagnetic results is also obvious from the polar wander curves (Fig.131) drawn for large continental regions such as Europe–North Asia and North America, when it is known that each of these regions consists of several distinct geological and tectonic units. In the author's opinion even an attempt to draw a simple polar wander curve for southern Europe would be a hazardous task in view of the complicated tectonics of the various small blocks comprising this land mass. This aspect is being increasingly realized by palaeomagnetists, and in more recent analyses (e.g. Deutsch, 1969; Hospers and Van Andel, 1970; McElhinny, 1973) the data are presented on a subcontinental scale in terms of stable platforms and blocks. However, with the increasing influx of palaeomagnetic data the number of blocks (or the so-called plates) is becoming inconveniently large. We shall revert to a discussion of continental drift in terms of plate movements in Chapter 8.

Palaeomagnetism and palaeolatitudes

An alternative method of presenting palaeomagnetic data is to draw a palaeographic map of the region in terms of ancient latitudes (using the formula $\tan I = 2 \tan \phi$, see p.167) and azimuths with respect to the geographic north (Fig.132). This approach, first used by Blackett et al. (1960), is particularly suitable for the comparison of palaeomagnetically determined latitudes with past geographical latitudes estimated from various palaeoclimatic indicators. There is spectacular evidence of Permo-Carboniferous glaciation having once occurred in areas which are now in low latitudes, e.g. India, Australia, South Africa, and South America. In particular Blackett (1961) has emphasized that palaeomagnetic observations are in agreement with the high palaeolatitudes implied by the widespread glaciation in these areas at those times.

Detailed comparisons of the palaeomagnetic data with other palaeoclimatic indicators are made by Irving (1964)

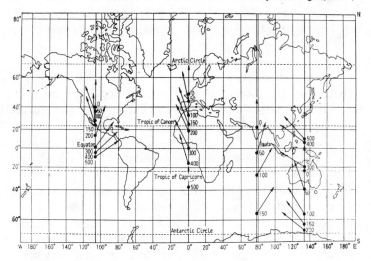

Fig.132. Ancient latitudes and orientations of North America, Europe, India, and Australia at various geological ages. The number against each point indicates its age in million years. (After Blackett et al., 1960.)

and Briden (1970b). The determination of palaeolatitudes by palaeomagnetic data has several applications which can be of statistical importance in the search for oil and mineral deposits. For instance, a majority of the Carboniferous coals of western Europe and North America fall in palaeolatitudes of less than 30°, whereas the Permian and younger coals from Canada, Siberia, and some southern continents fall in the high palaeolatitude group. Statistically, most large oil deposits in Tertiary beds are known to have formed within 30° of the palaeo-equator (Deutsch, 1965), and those occurring in Palaeozoic beds have palaeolatitudes of less than 20°. The palaeolatitude of source rocks is, therefore, of great significance in determining the potential accumulation of oil and natural gas. In the same way, the occurrence of certain evaporites and minerals such as gypsum, anhydrite, laterite, etc., is only possible in areas which were at one time in the appropriate low palaeolatitudes. For a more detailed account of palaeolatitude studies in relation to various palaeoclimatic indicators, the reader may refer to a collection of papers edited by Nairn (1964).

Archaeomagnetism and palaeointensity studies

Magnetic observatory records only reach back 400 years and this period of time is too short to allow a prediction of the long-term behaviour of the earth's magnetic field. The record has been extended by applying palaeomagnetic methods to dated archaeological material, to historically recorded lava flows and, with somewhat less precision, to dated varved sediments. Using the eleven well-dated historic lava flows of Mt. Etna, Chevallier (1925) traced the geomagnetic field variation from 394 B.C. to 1911. In extending the record to historic and prehistoric times, brick-built pottery kilns, tiles, pottery, varved clays, etc., have often been used, and in this way an offshoot of palaeomagnetism, archaeomagnetism, has been developed, based on the knowledge of secular variation of the geomagnetic field for the past few thousand years. Secular-variation studies of the direction of the field at various places for specific periods have been made by various workers in England, France, Japan, Sweden, the U.S.A., and the U.S.S.R. Nagata and Ozima (1967) have summarized recent data from various sources. The available data are so fragmentary that no definite conclusions can be drawn, although some of the data from Japan show an apparent periodicity of about 1000 years for a cyclic secular variation. In suitable cases the secular-variation curve can be used as a dating tool in archaeology and as a supplementary aid in the correlation of most recent lava flows.

Archaeomagnetism is also important when applied to archaeological prospecting. Buried archaeological objects, particularly burnt features (e.g. pottery kilns, fire places, etc.) exhibit a relatively strong TRM and give rise to local magnetic anomalies which can be detected by a ground magnetometer survey. Aitken (1961) and Abrahamsen (1967) have given examples of the magnetic location of archaeological objects.

Past variations in the intensity of the geomagnetic field have also been studied from archaeomagnetic measurements, particularly in France, Japan, and Czechoslovakia. The classic work in palaeointensity was first done by Thellier and Thellier (1959) on baked bricks and pottery of known archaeological age, and their method has been widely adopted as a standard technique. Bucha (1965) has recently summarized the results which indicate that the earth's field has decreased in intensity by a factor of about 1·5 over the past 2000 years. In a more recent analysis of all available historic and archaeological data, Smith (1970) has extended the palaeointensity record back to about 7000 B.C. (Fig.133); this shows an apparent decrease of the earth's dipole moment in the last 1500 years from a peak value about 50% higher than the present value. This rate of decrease, if persistent, may be indicative of an impending polarity reversal.

Applications to small-scale geologic problems

Palaeomagnetism is being increasingly applied to many

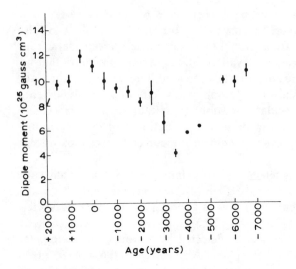

Fig.133.Variation in the geomagnetic dipole moment over the past few thousand years as deduced from palaeointensity data. Determinations have been averaged over 500-year intervals. (After Smith, 1970.)

types of local geological problems relating to stratigraphical, structural or tectonic investigations. The examples to be discussed here have been chosen primarily to illustrate the usefulness of palaeomagnetic studies in supplementing geological information.

(*1*) *Bending of Japan.* The Japanese islands have been subjected to numerous palaeomagnetic investigations. It has been shown by Kawai et al. (1961) that a consistent difference of about 53° in the mean directions (declination only) of NRM exists between the pre-Tertiary rocks from the northeast and southwest limbs of the Japan arc, whereas the Tertiary and Quaternary rocks from both limbs show directions which are close to the geographic north and also to each other (Fig. 134). According to these authors, the difference in declination is attributable to the main deformation at the middle to the extent of about 53° in the early Tertiary. This agrees with the

255

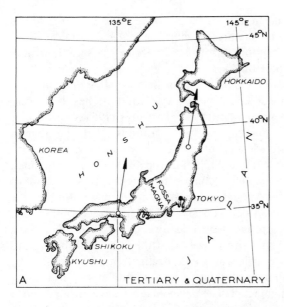

Fig.134.Bending of Japan illustrated by the remanent magnetization directions of pre-Tertiary and post-Tertiary rocks. (After Kawai et al., 1961; redrawn from Hospers and Van Andel, 1969.)

fact that the mean axes of the southwest and northeast limbs form an angle on the present map of about 123°, indicating an approximate bending through 57° about the middle of the main island of Honshu.

(2) *Rotation of Spain.* The idea that the Iberian peninsula has undergone a counter-clockwise rotation has been very popular among many "drifters" and Carey (1958) in particular proposed, on purely geological grounds, that the Iberian peninsula suffered an anti-clockwise rotation through 35° relative to stable (extra-Alpine) Europe in late Mesozoic times. Several palaeomagnetic studies relating to the rotation of Spain have been made, the most comprehensive study is the one made by Van der Voo (1969). After subjecting the rock samples to thorough magnetic and thermal cleaning tests, his results for that site showed that the Permo-Carboniferous samples yield a palaeomagnetic declination which differs by about 35° from that of stable Europe, the difference in inclination being insignificant. The palaeomagnetic evidence put forth by Van der Voo appears to confirm completely the anti-clockwise rotation envisaged by Carey. In this connection, it may be noted that Bullard et al. (1965), while attempting a "fit" of the northern continents, found that a clockwise rotation of Spain through 32° is required to close the Bay of Biscay at the 500-fathom isobath.

(3) *Relative rotation of the Bohemian Massif.* Birkenmajer et al. (1968) carried out palaeomagnetic studies of igneous and sedimentary rocks of Upper Carboniferous age from the Inner Sudetic basin and the Bohemian massif. They found a significant difference of about 17° in the palaeomagnetic declination for the two regions (Fig.135).

There is a complete accord between the results of the measurements on igneous and sedimentary rocks. Also, the differences in age, or in secondary remagnetization, do not appear to be able to explain the significant difference in palaeomagnetic declination between the two regions. Birkenmajer et al. (1968) therefore conclude that the Inner Sudetic basin and the Bohemian massif have suffered a relative post-Carboniferous rotation about a vertical axis through $17°$. Whereas the palaeomagnetic evidence for the rotation appears to be quite convincing, it is less easy to decide which area has rotated relative to Europe as a whole.

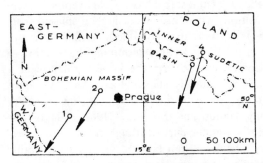

Fig.135.Sketch map showing the sampling areas in the Inner Sudetic basin (3 and 4) and the Bohemian massif (1 and 2) of Upper Carboniferous rocks. Also shown are the directions of the horizontal component of the remanent magnetization of the rocks studied. (Based on Birkenmajer et al., 1968; reproduced from Hospers and Van Andel, 1969.)

(4) Study of igneous intrusives. Several palaeomagnetic studies of igneous intrusive bodies have been made which have structural implications. Blundell and Read (1958) described a palaeomagnetic investigation of the younger gabbros of Aberdeenshire in northeast Scotland. Earlier it had been suggested by various workers that these gabbros (probably of early Palaeozoic age) suffered considerable relative tectonic movements since their consolidation as implied by tilted banded structures in the rocks. This suggestion could be tested by palaeomagnetism. By the comparison of the palaeomagnetic directions obtained for the separate gabbro masses, the authors concluded that the basic masses have remained relatively undisturbed since their formation, without undergoing any relative tectonic movements.

In another study, reported by Books et al. (1966), the average remanent magnetization of the gabbro complex that intrudes into the steeply dipping middle Keweenawan lavas (in northern Wisconsin) was measured to settle the dispute about the timing of the intrusion. The palaeomagnetic directions for the intrusive complex suggested that the intrusion of the gabbro occurred after tilting of the invaded lava flows.

(5) Palaeomagnetic dating and correlation. Rocks of increasing age from a subcontinental stable block yield average palaeomagnetic pole positions which are increasingly displaced from the present pole. It is, therefore, possible to estimate the age of remanence of a suite of rocks by determining the position of its average palaeomagnetic pole on the known polar wander curve for the same tectonically stable block. An example of palaeomagnetic dating of hydrothermal deposits in Czechoslovakia is given by Hanus and Miroslav (1963). The precise definition of a polar wander curve through reliable measurements is necessary before it can be used for magnetic dating of rocks belonging to the same tectonic unit. The potential application of palaeomagnetism as a dating tool is of special significance in the study of unfossiliferous sediments. The method has been applied by Athavale et al. (1972) to establish the age of some Precambrian sedimentary formations from India that are devoid of any material suitable for radiometric dating.

Distinct differences in palaeomagnetic directions obtained from rocks formed during major igneous events in an area can be of geological significance even when the polar wandering curve for the area has not been established. Significant differences in the directions of remanence imply age differences, although the magnitude of such differences cannot be assessed. For example, Precambrian dykes of similar orientation (the Abitibi swarm), in the Canadian shield, show three distinct groups of directions (Fig.136) and are, therefore, probably of three different ages (Larochelle, 1966). In a study, reported by Fahrig et al. (1965), of the Mackenzie dyke swarm (also in the Canadian shield), the direction of the samples collected from widely separated dykes, which vary in strike by at least $30°$, are found to be well grouped. This suggests that the intrusion of the swarm was an igneous event of comparatively short duration.

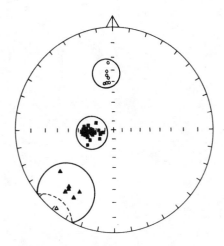

Fig.136.Relative dating by remanent magnetization directions. Dykes of a similar orientation in Canada, the Abitibi swarm, show three distinct groups of directions and are probably of three different ages. (After Larochelle, 1966; redrawn from Tarling, 1971.)

In certain cases the relative intensities of remanence can also be used to indicate age differences. Some suites of the Precambrian dykes of the Canadian shield have been classified and relatively dated by a detailed study of the Q ratios of a systematic collection (Strangway, 1966). The method has to be applied with due caution on rocks of similar composition and, in general, it cannot be applied to post-Precambrian times.

Some other examples of palaeomagnetism applied to correlation problems that can be mentioned are: the correlation of the Tertiary lava sequences, 15 km apart at Skye, Scotland (Khan, 1960); the widespread use of geomagnetic polarity reversals in correlating sections as an aid to field mapping in Iceland (Rutten and Wensink, 1960); the palaeomagnetic correlation of deep-sea sedimentary cores (Opdyke, 1968); and the dating of the North Sea Pleistocene sediments by palaeomagnetic correlation (Montfrans, 1971).

For more field examples and applications of palaeomagnetism to various geological and geophysical problems, the reader may refer to the texts by Irving (1964) and Tarling (1971).

CHAPTER 5
Earth Resistivity Methods

INTRODUCTION

There are a number of ways in which electric current can be employed to investigate subsurface conditions in an area. In the most commonly used method the current is driven through the ground using a pair of electrodes, and the resulting distribution of the potential in the ground is mapped by using another pair of electrodes connected to a sensitive voltmeter. From the magnitude of the current applied and from the knowledge of the current electrode separation it is possible to calculate the potential distribution and the path of the current flow if the underground were homogeneous. Anomalous conditions or inhomogeneities within the ground, such as electrically better or poorer conducting layers, are inferred from the fact that they deflect the current and distort the normal potentials. This is, in brief, the principle of measuring subsurface variation in electrical *resistivity* (reciprocal of conductivity) within the earth. In the 1920's the technique of the method was perfected by Conrad Schlumberger who conducted the first experiments in the fields of Normandy.

In practice, there are other complicated electrical effects which may create potentials other than that caused by simple ohmic conduction of the applied current. For example, electric potentials can be developed in the earth by electro-chemical actions between minerals and the solutions with which they are in contact. No external currents are needed to map the distribution of such spontaneous mineralization potentials. The detection of these potentials forms the basis of the "self-potential" method of prospecting for ore bodies such as pyrites. Moreover, electrical charges sometimes accumulate on the interfaces between certain minerals as a result of the flow of electric current from an external source. The method of "induced polarization" is based on this phenomenon in the search for metallic minerals and groundwater. In addition, slowly varying potentials are caused by natural (telluric) currents flowing inside the earth. These currents are believed to be induced inside the earth by ionospheric currents and, in general, they are capable of extending deep into the earth's crust. The "telluric method" was first applied in exploration geophysics during the 1930's and since then it has been extensively used in France, the U.S.S.R., and Africa.

A general introduction to various geoelectrical methods can be found in texts such as Dobrin's (1960) and Parasnis' (1972). The theory of electrical methods is relatively complicated and has been dealt with by Grant and West (1965), Bhattacharya and Patra (1968), and Keller and Frischknecht (1966). A discussion of the various electromagnetic methods used when searching for ore deposits will be found in Parasnis (1973). In this book we are primarily concerned with only the resistivity methods. The aim of this chapter is to describe the principles and practices of the resistivity methods, with some examples of the applications to geological problems. A brief discussion of the deep geoelectrical sounding methods as applied to crustal studies is also included.

RESISTIVITIES OF ROCKS AND MINERALS

The property of the electrical resistance of a material is usually expressed in terms of its resistivity. If the resistance between opposite faces of a conducting cylinder of length l and cross-sectional area A is R, the resistivity is expressed as:

$$\rho = RA/l \qquad (5.1)$$

The SI-unit of resistivity is ohm metre (Ωm). The conductivity σ ($=1/\rho$) of a material is defined as the reciprocal of its resistivity and measured in mho per metre, the word "mho" being coined by spelling "ohm" backwards.

The electrical conduction in most rocks is essentially electrolytic. This is because most mineral grains (except metallic ores and clay minerals) are insulators, electric conduction being through interstitial water in pores and fissures. Hence the resistivity of a rock formation generally depends on the resistivity of the contained electrolyte and is inversely related to the porosity and the degree of saturation. Generally speaking, hard rocks are bad conductors of electricity, but conduction may take place along cracks and fissures. In porous sedimentary formations, the degree of saturation and the nature of the pore-electrolytes governs the resistivity. In dry state most rocks are non-conducting.

Resistivity is, therefore, an extremely variable parameter, not only from formation to formation but even within a particular formation. There is no general correlation of the lithology with resistivity. Nevertheless, a broad classification is possible according to which clays and marls, sands and gravel, limestones and crystalline rocks stand in order of increasing resistivity. Fig.137 shows the approximate resistivity ranges of common rock types. More data on the resistivity of rocks and minerals (including ore deposits) can be found in Parasnis (1971).

FUNDAMENTALS OF THE CURRENT FLOW IN THE EARTH

Potential distribution in a homogeneous medium

The simplest approach to the theoretical study of the current flow in the earth is to consider first the case of a completely homogeneous isotropic earth layer of uniform resistivity. For a quantitative treatment, let us consider a homogeneous layer

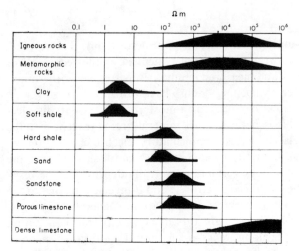

Fig.137.Approximate resistivity ranges of some rock types. The resistivity scale (horizontal) is logarithmic. (Adopted from Griffiths and King, 1965.)

of length *l* and resistance *R* through which a current, *I*, is flowing. The p.d. (potential difference) across the ends of the resistance is given by Ohm's law and is:

$$\Delta V = RI \qquad (5.2)$$

The resistance, *R*, of the layer is specified by its length, *l*, area of cross-section *A*, and the resistivity, ρ. By definition (eq.5.1), $R = \rho \, l/A$, and, therefore, eq.5.2 can be rewritten as:

$$\Delta V/l = \rho I/A \qquad (5.3)$$

or:

$$\text{grad } V = \rho i \qquad (5.4)$$

where grad *V* stands for the potential gradient, and *i* is the current density per unit of cross-sectional area.

The next step in the development of the theory is to derive the potential in a homogeneous medium due to a point source of the current. Now consider a semi-infinite conducting layer of uniform resistivity bounded by the ground surface and let a current of strength $+I$ enter at point C_1 on the ground surface (Fig.138). This current will flow away radially from the point of entry and at any instant its distribution will be uniform over a hemispherical surface of the underground of resistivity ρ.

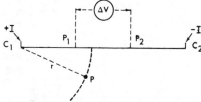

Fig.138.Method of calculating potential distribution due to a current source in a homogeneous medium (for explanation see text).

At a distance, r, away from the current source, the current density, i, would be:

$$i = I/2\pi r^2 \qquad (5.5)$$

The potential gradient $-\partial V/\partial r$ associated with the current is given by eq.5.4 which, when using eq.5.5, can be written as:

$$\frac{-\partial V}{\partial r} = \rho i = \rho I/2\pi r^2 \qquad (5.6)$$

The potential at distance r (e.g. at point P in Fig.138) is obtained by integrating eq.5.6 and is:

$$V = I\rho/2\pi r \qquad (5.7)$$

This is the basic equation which enables the calculation of the potential distribution in a homogeneous conducting semi-

266

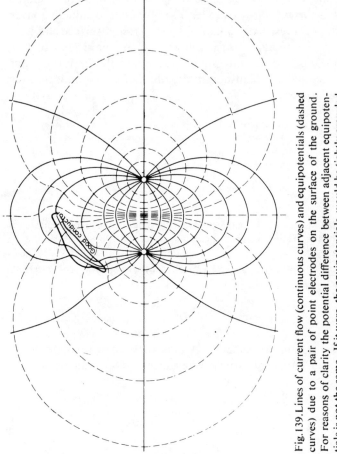

Fig.139. Lines of current flow (continuous curves) and equipotentials (dashed curves) due to a pair of point electrodes on the surface of the ground. For reasons of clarity the potential difference between adjacent equipotentials is not the same. If it were, the equipotentials would be tightly crowded near the electrodes and farther apart in the centre. The lower part of the figure representing the current distribution in a homogeneous ground is exact and not merely schematic. (After Parasnis, 1973.)

infinite medium. Fig.139 shows the distribution of the potential and the lines of current flow in the vertical section of a homogeneous conducting underground medium due to a pair of current electrodes. The figure also shows the distortion of the current flow lines caused by the presence of a good conductor.

From eq.5.7 it is easy to see that the p.d. between points P_1 and P_2 (Fig. 138) caused by current $+I$ at the "source" (entry point C_1) is:

$$V = \frac{I\rho}{2\pi} \left(\frac{1}{C_1 P_1} - \frac{1}{C_1 P_2} \right) \tag{5.8}$$

In the same manner, the p.d. between P_1 and P_2 caused by $-I$ current at the "sink" (exit point C_2) is:

$$V = \frac{-I\rho}{2\pi} \left(\frac{1}{C_2 P_1} - \frac{1}{C_2 P_2} \right) \tag{5.9}$$

The total p.d. between P_1 and P_2 is, therefore, given by the sum of the right-hand sides of eqs. 5.8 and 5.9, and is:

$$V = \frac{I\rho}{2\pi} \left(\frac{1}{C_1 P_1} - \frac{1}{C_1 P_2} - \frac{1}{C_2 P_1} + \frac{1}{C_2 P_2} \right) \tag{5.10}$$

or :

$$\rho = 2\pi \frac{\Delta V}{I} \frac{1}{G} \tag{5.11}$$

where G is an abbreviation for the expression in brackets in eq.5.10 and denotes the geometric factor of an electrode configuration.

The value of ρ determined in this way for a homogeneous conducting medium is independent of the positions of electrodes and is not affected when the positions of the current and potential electrodes are interchanged.

Apparent resistivity and true resistivity

Eq. 5.11 can be used to compute the true resistivity of the underground provided the underground is completely homogeneous. For inhomogeneous underground the resistivity as computed from eq.5.11, varies with the position of the electrodes, for example, if the current electrodes are moved while the potential electrodes are kept fixed. Also, if a given electrode configuration is moved as a whole, a different value of ρ would be obtained for each position of the array, provided lateral variations in resistivity exist within the ground. The resistivity obtained from eq.5.11 for an inhomogeneous underground is, therefore, designated as the apparent resistivity (ρ_a).

The apparent resistivity is a formal, rather artificial concept and it should not be considered to be some sort of average of resistivities encountered in the heterogeneous underground formation. Nevertheless, the concept is very useful in practical application of the resistivity method to subsurface investigations. For a proper interpretation of this quantity (ρ_a) one must always bear in mind the configuration with which it has been determined.

Potential and current distribution across a boundary

At the boundary between two media of different resistivities the potential remains continuous while the current lines are refracted according to the law of tangents as they pass through the boundary. With the notation of Fig.140, the law of the refraction of current lines can be written as:

$$\rho_1 \tan \alpha_1 = \rho_2 \tan \alpha_2 \qquad (5.12)$$

(An elementary derivation of this equation can be found in Griffiths and King (1965, p. 15).)

If $\rho_2 < \rho_1$ the current lines will be refracted away from the normal. The result of the refraction of the current lines is that they would diverge or converge appreciably from the normal radial pattern (cf. Fig.139) as they approach the boundary.

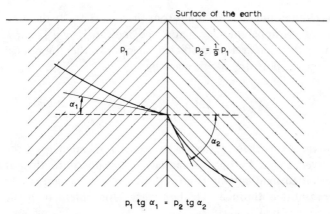

Fig.140.Refraction of current lines crossing a boundary between two media of different resistivities. (After Kunetz, 1966.)

The refraction of current lines is not the only effect of a boundary separating two electric media. The overall potential distribution and the current flow in the medium around the source is also affected by the boundary. A simple case of two layers can be solved by the method of electrical images.

Consider a source of current, I, at point S in the first layer (ρ_1) of semi-infinite extent bounded by a horizontal ground surface on the top (Fig.141) and a vertical boundary on the right which separates it from another layer of resistivity ρ_2.

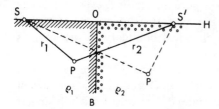

Fig.141.The electrical image method of calculating the potential distribution in two media of different resistivities (ρ_1, ρ_2). H is the ground surface and B is the vertical boundary (for explanation see text).

By optical analogy the potential at any point P would be that from S plus the amount reflected by the layer (ρ_2), as if the reflected amount were coming from the image S'. If the dimming of the apparent source at S' be indicated by a factor k (similar to the reflection coefficient in optics), then by optical analogy the potential at P is given by:

$$V_{1\ (P)} = \frac{I\rho_1}{2\pi} \left(\frac{1}{r_1} + \frac{k}{r_2} \right) \tag{5.13}$$

In the case where P lies in the second medium (ρ_2), the observed effect, by optical analogy, corresponds to that of transmitted light coming from S. Since only a fraction $(1-k)$ is transmitted through the boundary, the potential in the second medium is given by :

$$V_{2(P)} = \frac{I\rho_2}{2\pi} \left(\frac{1-k}{r_1} \right) \tag{5.14}$$

Continuity of the potential requires that at the boundary, where $r_1 = r_2$, $V_{1(P)}$ must be equal to $V_{2(P)}$. Therefore, equating 5.13 and 5.14 we obtain:

$$\frac{\rho_1}{\rho_2} = \frac{1-k}{1+k}, \text{ or } k = \frac{\rho_2 - \rho_1}{\rho_2 + \rho_1} \tag{5.15}$$

The value of the dimming factor, k, always lies between ± 1; if the second layer is a pure insulator $(\rho_2 = \infty)$, $k = +1$; and if the second layer is a perfect conductor $(\rho_2 = 0)$ then $k = -1$. When $\rho_2 = \rho_1$ no electrical boundary exists, and, as might be expected, $k = 0$.

In the above treatment, eqs. 5.13 and 5.14 are valid for a single point source of current $+I$. The effect of a $-I$ current source (second current electrode) can be calculated in a similar manner and the total potential in each medium can be determined by superposition.

Now imagine that the conditions shown in Fig.141 are modified. If the boundary separating the two layers is horizontal (instead of vertical), the problem is complicated by the fact that there are now three layers, the air forming the third layer. In these situations an infinite number of "images" is produced of a source placed at the ground surface. The situation is somewhat analogous to that of a light source (e.g. a candle) placed between two parallel mirrors. The resulting potential at, for instance, a point on the ground surface is the sum of the potentials due to the current source and its infinite series of images. The expression for the potential, though quite long, converges in the calculation. The complications in computation increase with the increase in the number of layers, and for a multi-layer case use of a computer is almost essential.

The details of the method of computation of theoretical ρ_a curves for a multi-layered earth structure are beyond the scope of this book. The interested reader is referred to papers by Flathe (1955) and Van Dam (1967). Master curves of ρ_a for two-, three-, and four-layer cases for the Schlumberger configuration can be found in Orellana and Mooney (1966); this work also includes master tables of ρ_a for the Wenner configuration.

ELECTRODE ARRANGEMENTS AND FIELD PROCEDURE

Electrode configurations

For field practice a number of different electrode configurations have been proposed. Several commonly used linear array-type arrangements are shown in Fig.142.

In the Wenner array four electrodes are equally spaced along a straight line. The distance between adjacent electrodes is called the array spacing, *a*. For this configuration the equation of (apparent) resistivity (eq. 5.11) reduces to:

$$\rho_a = 2\pi a \ \frac{\Delta V}{I} \qquad (5.16)$$

In the generalized Schlumberger array the distance (2*l*) between the potential electrodes is small compared to the dis-

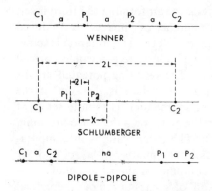

Fig.142. Some commonly used electrode configurations in resistivity survey-ing. C_1, C_2, and P_1, P_2 denote the positions of the current and potential electrodes.

tance $(2L)$ between the current electrodes. If $L \geqslant 5l$, the equation of resistivity (eq.5.11) approximates to:

$$\rho_a \rightleftharpoons \frac{\pi}{2l} \frac{(L^2 - x^2)^2}{L^2 + x^2} \frac{\Delta V}{I} \tag{5.17}$$

where x is the distance of the observation point (the point mid-way between the potential probes) from the centre of the line $C_1 C_2$.

In the symmetrical Schlumberger array arrangement, $x=0$, and the formula for resistivity simplifies to:

$$\rho_a \simeq \frac{\pi L^2}{2l} \frac{\Delta V}{I} = \frac{\pi L^2}{I} \frac{\Delta V}{\Delta r} \tag{5.18}$$

where, obviously, $\Delta V/\Delta r$ is the surface gradient of potential, i.e. the electric field, E, at the observation point.

In the so-called dipole-dipole configuration the potential probes, $P_1 P_2$, are outside the current electrodes, $C_1 C_2$, each pair having a constant mutual separation, a. If the distance between the two pairs, na, is relatively large, the current source may be treated as an electric dipole (analogous to a magnetic dipole). The equation of resistivity then becomes:

$$\rho_a \simeq \pi n\,(n+1)\,(n+2)\,a\,\frac{\Delta V}{I} \qquad (5.19)$$

Non-collinear dipole arrangements have also been used for deep electrical soundings, especially in the U.S.S.R.

Although there are other possible configurations, some using less than four electrodes, they are not commonly used. The Wenner and the Schlumberger arrays are by far the two most widely used setups.

The aim of the resistivity survey is to delineate resistivity boundaries (both horizontal and vertical) in a heterogeneous ground. In practice this is accomplished by two distinct procedures often called, by analogy, electric sounding (or drilling) and electric profiling (or trenching).

Electric sounding (or drilling)

When the ground consists of a number of more or less horizontal layers, knowledge of the vertical variation in resistivity is required. The object of electric drilling is to deduce the variation of resistivity with depth below a given point on the ground surface, and to correlate it with the available geological information in order to infer the depths and resistivities of the layers (formations) present. The procedure is based on the fact that the current penetrates continuously deeper with the increasing separation of the current electrodes. Fig.143 illustrates the concept as applied to a two-layer problem. When the electrode separation, $C_1 C_2$, is small compared with the thickness, h, of the upper layer, the apparent resistivity as determined by measuring ΔV between the potential electrodes, $P_1 P_2$, would be virtually the same as the resistivity, ρ_1, of the upper layer. This is because a very small fraction of the current

274

would penetrate in the substratum below the boundary. As the electrode separation is increased a greater fraction of current will penetrate deeper, the lines of current flow being distorted at the boundary. At spacings, which are very large compared with h, the apparent resistivity approaches ρ_2 because the fraction of current confined to the surface layer becomes negligible.

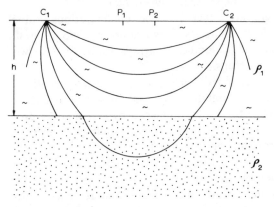

Fig.143.Principle of electric drilling. For small current-electrode separation ($C_1C_2 < h$) the current is virtually confined to the surface layer (ρ_1). As the separation C_1C_2 increases a greater fraction of the current penetrates deeper in the substratum (ρ_2).

Fig.144 shows typical two- and three-layer curves for the variation in apparent resistivity as a function of the current electrode separation for the symmetrical Schlumberger electric sounding in which the potential probes are kept fixed and the current electrodes are systematically moved outwards in steps.

In electric sounding with the Wenner configuration (see Fig.142) the array spacing, a, is increased by steps, keeping the midpoint of the configuration ("drilling point") fixed. A typical set of electrode separations is $a = 2, 6, 18, 54, \ldots$ m. The resistivity curves (ρ_a vs. a) for Wenner soundings, although of the same general form, are different from that for Schlumberger soundings.

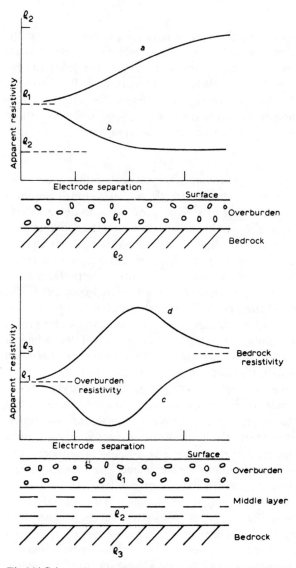

Fig.144. Schematic electric sounding curves over a horizontally stratified earth. The apparent resistivity (ρ_a) curves are shown for the following cases: (a) $\rho_2 > \rho_1$, (b) $\rho_1 > \rho_2$, (c) $\rho_3 > \rho_1 > \rho_2$, and (d) $\rho_2 > \rho_3 > \rho_1$. (After Parasnis, 1973.)

The present trend in resistivity soundings favours the Schlumberger configuration. Since only two electrodes are moved, it is much more convenient to use the latter for the field routine than the Wenner configuration. Furthermore, as the potential electrodes remain fixed, the effect of near-surface inhomogeneities in their vicinity (due to soil condition, weathering, etc.) is constant for all observations.

Electric profiling

If the layers or boundaries are vertical, rather than horizontal, planes another procedure (the so-called electric profiling) is adopted. The object of electric profiling is to detect lateral variations in the resistivity of the ground. In the Schlumberger method of electric profiling (see Fig.145A) the current electrodes remain fixed at a relatively large distance, for instance, a few hundred metres, and the potential electrodes with a small constant separation (P_1P_2) are moved between C_1C_2. ρ_a is calculated from eq. 5.17 for each position that the mobile pair of potential electrodes takes. At the end of the profile line the Schlumberger setup is transferred on the adjacent line and so on, until the area to be investigated has been covered. In fact, the profile lines will usually be at right angles to the "strike" of the structures (e.g. faults or dykes) to be mapped so that we may expect to find somewhat similar results from profile to profile. In the Wenner procedure of electric profiling (Fig.145B) the four-electrode configuration with a definite array spacing a, is moved as a whole in suitable steps, say, 10–20 m along a line of measurement. The choice of array, spacing a, would primarily depend on the depth of the anomalous resistivity feature(s) to be mapped. The two curves in Fig.145 show the apparent resistivity curves obtained by Schlumberger and Wenner profiling across a vertical contact between two rock formations. The Wenner curve differs in that it has four cusps; the cusps may, however, not be observed in practice unless the measurements are taken at very close intervals.

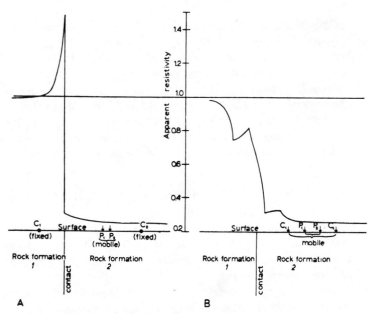

Fig.145.Apparent resistivity across a vertical contact between two rock formations using different electrode configurations. Formation 1 is four times as resistive as formation 2. (After Parasnis, 1973.)

Steep gradients in the resistivity curve are characteristic markers of structures with near-vertical boundaries, such as faults, dykes, and veins. For all practical purposes, a boundary that is inclined as much as 30° from vertical would yield essentially the same results.

RESISTIVITY SURVEY INSTRUMENTS

In the field, resistivity surveys can be carried out with simple equipment consisting of a high tension battery pack as the source of current, four metal stakes, a milliammeter, a voltmeter, and four reels of insulated cable. Most of the field measurements are nowadays made by using alternating current which offers definite advantages. Advantages include the ease of power production, the facility to amplify measured potentials, and above all the ability to avoid unwanted potentials such as caused by polarization of electrodes or by natural earth currents (telluric currents). However, it is very desirable that a low-frequency alternating current be used otherwise, because of "skin effect", there would be a rapid decrease of the current intensity with depth, and consequently a decreased depth of investigation. For an alternating current the so-called "depth of penetration", h, is given by the expression:

$$h = 503 \cdot 8 \sqrt{\rho/n} \tag{5.20}$$

where h is the depth in metres at which the current density is reduced to $1/e$ ($= 1/2 \cdot 718$) of its value at the surface, ρ the resistivity (Ωm) of the ground, and n the frequency of the alternating current used.

One of the most commonly used portable instruments is the Swedish-make Terrameter (Fig.146) which operates at a low frequency of about 4 c/s and is suitable for moderate depths of investigation (in favourable conditions down to about 100 m). Its maximum output is 6 W and it covers a resistance range from 0·01 to 10·000 Ω; through the use of a calibrated potentiometer the quantity $\Delta V/I$ (which is needed for calculating ρ_a from eq. 5.11) is directly read as the resistance, R.

APPLICATION AND INTERPRETATION OF RESISTIVITY DATA

The data of resistivity surveys can be presented in two forms, profiles and maps, for the purpose of geological interpretation.

In the case of electric sounding with an expanding electrode configuration it is usual to present the results as a series of graphs (curves) expressing the variation of ρ_a with increasing electrode separation. These curves represent, at least qualitatively, the variation of resistivity with depth. In relatively simple cases involving only two or three horizontal layers,

Fig.146.The ABEM Terrameter for resistivity measurements. (Courtesy of ABEM Company, Stockholm.)

estimates of the depths to the interfaces can be made and, in the light of the available geological information, a fairly satisfactory picture of the stratification can often be deduced.

In the case of electric profiling with a constant electrode spacing, the data may be presented as graphs showing the resistivity variation along the traverses, or as a contour map showing the lines of equal resistivity. Such a map is an expression of the lateral resistivity variation of the ground in the range of the depth which is penetrable with a chosen electrode spacing. Usually the contour maps are interpreted qualitatively (by inspection) to delineate the zones of low and high resistivity of the expected geological formations over the range of the depth investigated. Plotting of contour maps may not always be essential. Two-dimensional vertical structures such as faults, dykes, contact zones, etc., are often easily detectable on the resistivity profiles measured across the strikes of these features.

The quantitative interpretation of resistivity data is one of the most intricate problems and the reader should constantly guard against simple rules of thumb in this respect. In spite of the elaborate mathematical study of the problem made by several authors (Hummel, 1932; Pekeris, 1940; Tagg, 1964; Zohdy, 1965; Koefoed, 1968), it is very difficult to obtain reliable results by applying a theoretical analysis to the resistivity data obtained in the field. This is because the theory developed so far can only be applied to simple plane-layered models, whereas in practice the variations in resistivity are usually much more complex both in lateral and vertical directions. Notwithstanding these difficulties, some geological situations can be approximated quite closely by simple-layered structures for which interpretation techniques based on the use of standard theoretical curves are applicable. It is possible to discuss here only some of these techniques and to present a few examples of their application to simple structural problems.

Mapping of layered horizontal structures

Two-layered stratifications may be of two types: (1) $\rho_2 > \rho_1$ (e.g. unconsolidated overburden lying on a bedrock); (2) $\rho_1 > \rho_2$ (e.g. poorer conducting alluvium lying over a better conducting sand or a clay formation). In either case the following procedure may be adopted for interpreting a Wenner sounding curve:

(a) Plot the observed apparent resistivity curve on a transparent double logarithmic sheet with ρ_a on the ordinate (y-axis) and a (the array spacing in Wenner configuration) along the abscissa (x-axis).

(b) Superpose the observed curve of ρ_a on the sets of two-layer master curves (see Fig.147), the upper set for case (1) and the lower set for case (2). Shift the field curve, keeping the axes parallel, until a good match is obtained.

(c) The origin of coordinates for the master curves ($a/h_1 = 1$, $\rho_a/\rho_1 = 1$) is known as the "index cross". This point corresponds to a point on the field curve with corresponding values for ρ_a and a. In the example shown (Fig.147), the coordinates of the index cross as read on the field plot yield $h_1 = 12$ m and $\rho_1 = 130$ Ω m.

(d) The value of k corresponding to the field curve is read from the master curve. If no good match is obtained the value of k is obtained by interpolation ($k \approx 0.45$ in Fig.147) between two adjoining master curves. Using the relation, $\rho_2/\rho_1 = (1 + k)/(1 - k)$, ρ_2 is calculated to be about 50 Ωm.

If the field curve of ρ_a is obtained by the Schlumberger sounding method, the interpretation procedure is essentially the same although curve matching has to be done with the set of standard Schlumberger sounding curves (Compagnie Générale de Géophysique, 1963; Orellana and Mooney, 1966).

The problem of interpretation of three-layer curves is slightly more complicated because of the increased number of parameters (ρ_1, ρ_2, ρ_3, h_1 and h_2). The situation is illustrated schematically in Fig. 144, curves c and d. As expected, ρ_a for a very small electrode separation would correspond roughly to the resistivity of top layer ρ_1. As the electrode separation is increased a significant part of the current enters the second layer and ρ_a would increase or decrease smoothly depending on whether ρ_2 is greater or less than ρ_1. With increasing electrode separation more and more current enters the third layer and for very large electrode separations ρ_a approaches the resistivity of the third layer (ρ_3).

The interpretation procedure for three-layer curves is more or less similar to that for two-layer curves. In short, the procedure consists of matching the left-hand part of the three-layer curve with a two-layer master curve. This enables the determination of ρ_2; ρ_1 and ρ_3 being determined by the ρ_a values for very small and very large electrode separations. When $\rho_1 : \rho_2 : \rho_3$ is known, the corresponding master curve sheet can be selected from the album of three-layer master curves, and by matching this with a standard curve the depth parameters are determined.

In passing it may be mentioned that there are more varieties of three-layer curves (other than those shown in Fig.144), for example, the "double ascending type" (when $\rho_1 < \rho_2 < \rho_3$), or the "double descending type" (when $\rho_1 > \rho_2 > \rho_3$). A more detailed discussion of three-layer curves will be found in Bhattacharya and Patra (1968).

Ambiguity in resistivity interpretation

The field example illustrated in Fig.147 indicates that stratigraphic information about ground layering can be obtained by resistivity mapping under favourable conditions. However, in general the resolving power of the method is not very high. This is particularly true for deeper boundaries, where substantial variations in resistivity must occur before their effects can clearly be distinguished from irregularities due to near-surface inhomogeneities.

In addition, the "principle of equivalence" and the "principle of suppression" introduce other types of ambiguity in the interpretation. For example, a relatively thin conductive layer sandwiched between two layers of higher resistivity will tend to concentrate current flow in it. The total current carried by it will be unaltered if we increase its resistivity, ρ, but at the same time increase its thickness, h, so that the ratio h/ρ is constant. Fig.148 illustrates this. On the other hand, a resistant bed sandwiched between two more conductive beds is characterized by the product of its thickness and resistivity. Thus, in this case, all middle layers for which the product $h\rho$ is constant are electrically equivalent. In either case, a unique determination of h and ρ would be difficult if not impossible. Also, a middle layer, with a resistivity intermediate between the resistivities of the enclosing beds, will practically have no influence

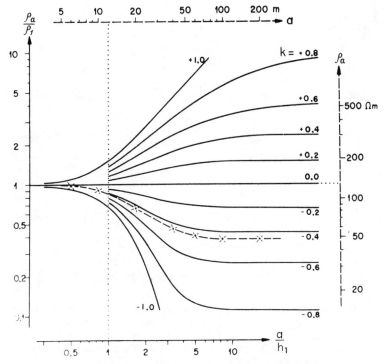

Fig.147.Example of the interpretation of a field curve (dashed line with crosses) by matching it with a set of standard two-layer resistivity curves. (After Gassmann and Weber, 1960.)

on the resistivity curve, as long as its thickness (in comparison with the depth) is not great enough. For small thickness its effect on the ρ_a curve is inappreciable so that the presence of the layer will be suppressed. This failure to detect beds of intermediate resistivity is found frequently in groundwater studies when a layer of wet alluvium is enclosed between dry alluvium (surface layer) and a shaly substratum.

Mapping of vertical structures

In contrast with the study of the effects of horizontal layers in the previous section, the effects of vertical structures (e.g. faults, fissures, dykes, veins, and shear zones) are lateral. If

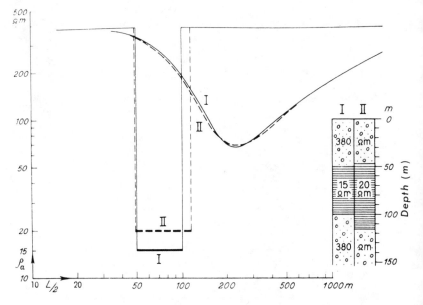

Fig.148.Illustration of the principle of equivalence for a conductive bed lying between two resistive beds. In the example shown, the resistivity curves are practically the same for two situations: (I) middle layer of resistivity 15 Ωm and thickness 50 m, or (II) resistivity 20 Ωm and thickness 66 m. (After Bentz, 1961.)

these features crop out, abrupt discontinuities in the slope of the ρ_a curves are obtained (see Fig.145) as the mobile electrode configuration crosses the vertical resistivity boundary. In practice, however, the sharp peaks will usually be somewhat subdued owing to the effect of top soil or moraine cover.

A vertical fault is one of the most useful geologic structures to study. Many of the key features found in the resistivity anomalies over a vertical fault are found also in anomalies over other near-vertical structures. In so far as the resistivity anomaly is concerned, the fault represents a vertical contact problem between two media of differing resistivity. Lögn (1954) has theoretically calculated apparent resistivity profiles across a vertical contact with different resistivity contrasts.

The curves of ρ_a in Fig.149 are for the case of a modified Schlumberger configuration where one of the current electrodes was set far away (theoretically at an infinite distance) from the mobile pair of potential electrodes P_1P_2. The vertical discontinuity is evident as a steep gradient in the resistivity curve at the contact. Fig.150 shows the field curve of ρ_a obtained

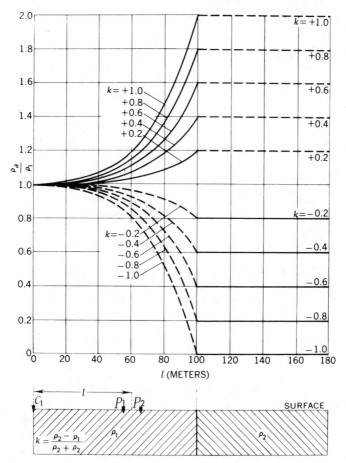

Fig.149.Horizontal resistivity profiles across a vertical fault for different resistivity contrasts—modified Schlumberger configuration. (Adopted from Lögn, 1954.)

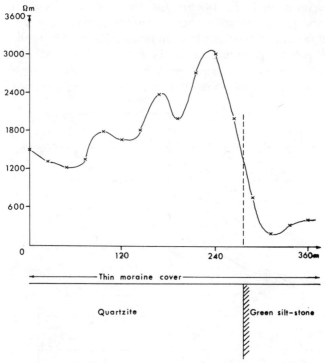

Fig.150.Apparent resistivity profile across a vertical contact between quartzite and green siltstone in southern Bornholm. The profile was measured with a mobile Wenner setup with a fixed array spacing, a, of 12 m.

by horizontal profiling with the Wenner configuration across a near-vertical contact between Balka quartzite and green siltstone in southern Bornholm. The effect of the varying thickness of the moraine cover overlying the two formations is apparent from the curve (cf. Fig.145).

The interpretation of resistivity data observed over "double contacts" is important for delineating features such as dykes, veins, and brecciated zones, all of which may be directly or indirectly related to mineral deposits. Fig.151 shows an observed resistivity profile across a shear zone and limestone fault block using the Wenner configuration. The true resistivity of the sheer zone is higher than that of the surrounding rock (sandstone). The ρ_a curve over the limestone block has a character similar to that over a wide dyke, whereas the ρ_a curve over the shear zone is similar to that over a thin resistant dyke.

Interpretation of resistivity maps

Resistivity maps are often made to obtain a general picture of the near-surface and subsurface conditions of formations in an area. In contrast with the resistivity profiles, the interpretation of resistivity maps is largely qualitative.

Fig.152 shows an example of the resistivity map of a part of the Kabul basin in East Afghanistan. The map summarizes the results of a series of Schlumberger depth soundings made with sets of varying electrode spacings from 400 to 1400 m.

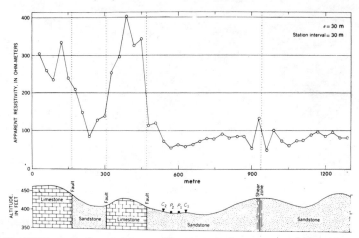

Fig.151.Horizontal resistivity profile across a shear zone and limestone fault block in Illinois—Wenner configuration. (After Hubbert, 1932.)

Within the basin sediments the deposits of relatively high specific resistivity could be localized. The extensive formation with a resistivity of about 250 Ωm represents a potentially

288

useful aquifer. Near the eastern border, below the present
valley of the Kabul river, underground rocks show strikingly
high resistivities. In light of the geological knowledge of the
area, such high resistivities indicate a strong cementation from
gravels and sand to conglomerates (almost non-porous). Drill-
ings repeatedly penetrate these high resistive formations. Homilius (1969) has discussed in detail the hydrogeological aspects
of the geoelectric resistivity measurements in East Afghanistan.
The other example is taken from a paper by Van Dam and

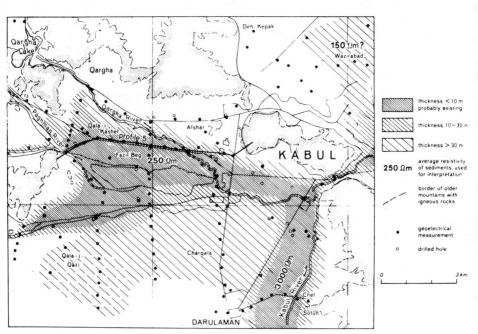

Fig.152.Zones of high resistivities in the Upper Kabul basin. (After Homilius, 1969.)

Meulenkamp (1967). Fig.153 shows the resistivity map of a
polder area in the province of Noord Holland. The primary
object of the resistivity measurements was to study the salinity
distribution of groundwater in the polder area and its environ-

ments. Interpretation of the measurements indicated the presence of fresh water on the east and west margins of the polder, whereas in the polder itself 30 m of the saline water appeared to lie over some tens of metres of fresh groundwater. This "reverse" salinity distribution is attributed to the presence of an impermeable layer of clay of Holocene age. The presence of such a clay layer was confirmed by some borelogs of the area.

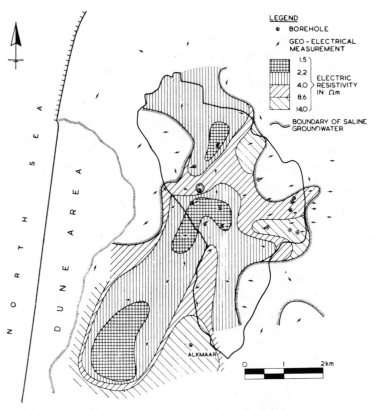

Fig.153.Resistivities of the sand layer with a saline groundwater boundary as determined by geoelectrical measurements in the province of Noord Holland, The Netherlands. (After Van Dam and Meulenkamp, 1967.)

Resistivity surveys have found wide applications in shallow-depth investigations relating to civil engineering and geo-technical problems. An interesting account of a survey for resistivity mapping of postglacial basins in Denmark has been reported by Klitten (1972).

GEOELECTRICAL DEEP SOUNDINGS

In the conventional method of resistivity sounding one measures the potential drop on the surface caused by an externally applied current. The depth of penetration obtained with the method depends mainly on the separation of the current electrodes. Application of the method in order to obtain information to depths of some thousands of metres would require impractically long current lines, apart from a large power supply. In practice, this limits the use of the conventional method for sounding to a depth of at most some hundreds of metres.

At present, most of the deep electric prospecting (especially in the U.S.S.R.) is being done by dipole sounding (see p. 273). It is claimed that when the length of the power dipole is 1000–1500 m, that of the measuring dipole 300–500 m, and the distance between them 10–12 km, it is possible to trace the basement at a depth of 2·5–3 km. Another great advantage of the dipole method is the possibility for sounding on a curved profile. For conventional sounding, the electrodes C_1, C_2, P_1, and P_2 must lie on a straight line. In difficult areas (e.g. forests and swamps) it is not always possible to maintain a straight profile for long distances. In dipole sounding, it is not necessary for the centres of all the dipoles to lie on the same straight line. More details about the dipole sounding will be found in Alpin et al. (1966).

Magneto-telluric sounding is another method of studying resistivities to great depths within the earth. The method is based on the fact that natural currents (believed to be induced in the earth by fluctuating ionospheric currents) flow in the earth and may well extend deep below the crust. The current distribution will, of course, depend on the resistivity of the rocks encountered. Because of the various astrophysical phenomena, telluric current fields are constantly fluctuating with time, in both direction and magnitude. This imposes the use only of comparative measurements with respect to a fixed reference point; measurements to be compared must be made at the same instant. Since the depth to which an electromagnetic field penetrates in a conductor depends on both the resistivity of the conductor and the frequency (see eq. 5.20), the resistivity may be computed as a function of depth within the earth if the amplitudes of the magnetic and electric field changes can be measured at several frequencies. A good review of the theoretical work and a summary of the worldwide results of magnetotelluric soundings has been given by Keller (1971).

The theory and instrumental details of deep-resistivity sounding methods are beyond the scope of this book. All that can be done here is to present a few examples of the application to crustal studies.

Deep sounding in the Rhine Graben

This example is taken from a paper by Blohm and Flathe (1970). Deep geoelectric sounding was carried out, with a maximum electrode distance C_1C_2 of 150 km, along the graben between Karlsruhe and Säckingen as part of the Upper Mantle Project programme. The basis of the operation was a Schlumberger arrangement with the recording centre midway between the high-voltage line Karlsruhe–Säckingen.

Fig.154 shows the sounding curve with a possible interpretation model. The resistivity distribution is shown in the upper part of the figure. It provides a clear illustration of the good conductivity of the graben fill ($\lesssim 10 \, \Omega$ m). This good conductivity, which was to be expected, screens the effects from the deeper underground in a decisive way and thus prevents any detailed information being obtained about the conditions in the basement.

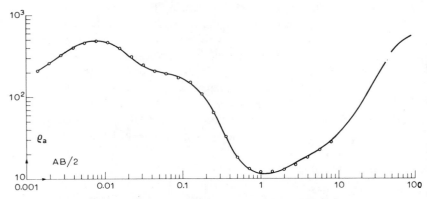

Fig.154. Comparison of the measured resistivity curve (solid line) in the Rhine Graben with the theoretical curve (open circles). The upper part of the figure shows the interpretation model with the true resistivities. (After Blohm and Flathe, 1970.)

Electrical model of the Icelandic crust

In recent years studies of the Icelandic crust have assumed great importance in connection with the understanding of the regional tectonics and thermal processes beneath the island. It is apparent that an important aspect of the large regional heat flow is the hydrothermal activity which is evident not only in the near-surface steam fields and hot springs but also in the unusually low resistivities at shallow depths in the crust.

From the results of a series of Schlumberger, dipole and magneto-telluric soundings Hermance (1973) has proposed a model (Fig.155) for the average electrical properties of the sub-Icelandic crust. Hermance's interpretation of the field resistivity data suggests: (1) the suppression of resistivity at shallow depths to values of 10–100 Ωm is caused by regional hydrothermal activity; and (2) appreciable effects from water may be present to depths of 8–10 km.

For more details on the deep electrical structure of Iceland the reader is referred to papers by Hermance and Garland (1968a, b).

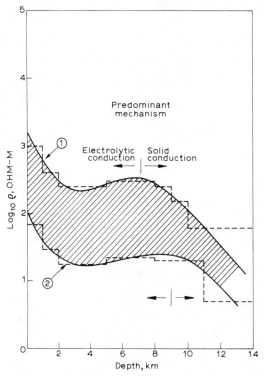

Fig.155. An electrical model for the sub-Icelandic crust. Curves 1 and 2 are based on the measured magneto-telluric response. The dashed lines are discrete layers of varying resistivities. The depths are indicated at which electrolytic conduction is dominated by solid conduction. (After Hermance, 1973.)

Magneto-telluric studies of sedimentary basins

An example of deep sounding for determining the resistivity distribution in sediments below the continental shelf has been given by Srivastava et al. (1973). Fig.156 shows the interpre-

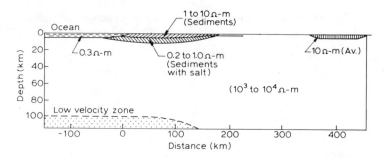

Fig.156.Simplified conductivity structure across the Nova Scotia continental shelf and Bay of Fundy as inferred from the magneto-telluric studies. (After Srivastava et al., 1973.)

tation of the results in terms of a simplified conductivity structure across the Nova Scotia continental shelf. The low resistivity under the shelf is probably associated with a highly saline interstitial fluid in the bottom part of the sedimentary section. At the bottom of the section a high concentration of evaporites may be responsible for the frequent salt domes or diapirs that are present on the shelf. A sediment thickness of several kilometres with a resistivity of less than 0.5 Ωm is required. Magneto-telluric measurements in the North German basin (Vozoff and Swift, 1968) also indicate sedimentary sections of a similarly low resistivity ($\sim$ 1 Ωm).

CHAPTER 6

Radiometric Methods

INTRODUCTION

Before the discovery of radioactivity (H. Becquerel, 1896), Kelvin and most other physicists held the view that the earth was only a few tens of millions of years old, while the geologists believed its age was a few hundreds of millions of years. The debate about the age of the earth would probably not have been settled today, if the radioactivity of rocks had not provided the decisive answer.

In many respects the study of the radioactivity of rocks and minerals is of great importance in geology and geophysics. First, the heat produced by radioactive disintegrations is probably the most important factor in the establishment of thermal conditions within the earth. Secondly, the rate of radioactive decay of certain naturally occurring elements in rocks provides a powerful means of dating geological events, in particular the times of formation of rocks in the earth's crust. In addition, since almost all rocks contain traces of radioactive elements in varying proportions, radiometric surveys can be used with advantage in prospecting for radioactive ore bodies.

This chapter is aimed to provide a brief discussion of the radioactive methods and their applications to problems of geochronology and geophysical prospecting.

FUNDAMENTALS OF RADIOACTIVE DISINTEGRATION

By the phenomenon of radioactivity we mean the disintegration of an atomic nucleus by emission of energy and particles of mass. The disintegrating nucleus ^AX_Z is transformed into a nucleus of another element with a change in atomic mass A and atomic number Z, the by-products of disintegration being α particles (helium nuclei $^4He_2$), β particles (electrons) and γ radiation in various combinations.

Radioactive disintegration or decay is a random process and is expressed in terms of the probability that a constituent particle in a nucleus will escape through the potential barrier binding it to the nucleus. The probability of decay is unaffected by physical conditions, like pressure and temperature, and depends upon the number of atoms present. It follows that the rate of decay of N nuclei of a particular species is directly proportional to N:

$$\frac{dN}{dt} = -\lambda N \qquad (6.1)$$

of which the solution is:

$$N = N_0 e^{-\lambda t} \qquad (6.2)$$

where N_0 is the number present at time $t = 0$. The factor λ, called the decay constant, is a unique property of each disintegrating nucleus.

The rate of decay is often quoted in terms of a related quantity, the half-life. This is the time taken to reduce the number of parent atoms by one half:

$$T = \ln 2/\lambda = 0 \cdot 69325/\lambda \qquad (6.3)$$

Decay mechanisms and experimentally determined decay constants (λ) of important radioactive elements in the earth are shown in Table VIII. If λ is known, measurement of the relative abundance of a parent isotope and the end or daughter product permits the determination of t from eq. 6.2. This is the

TABLE VIII

Decay schemes and decay constants of important radioactive elements in the earth
(reproduced from Stacey, 1969, p.224)

Parent isotope	Percent of natural element	Decay mechanism	Stable daughter	Decay constant (yr^{-1})	Half-life (yr)
^{238}U	99·274	$(8\alpha - 6\beta)$ series decay	^{206}Pb	$1\cdot54 \times 10^{-10}$	$4\cdot51 \times 10^{9}$
		$4\cdot5 \times 10^{-5}\%$ spontaneous fission	various	$6\cdot9 \times 10^{-17}$	
^{235}U	0·720	$(7\alpha - 4\beta)$ series decay	^{207}Pb	$9\cdot72 \times 10^{-10}$	$7\cdot13 \times 10^{8}$
		neutron-induced fission	various	very small (proportional to neutron × flux)	
^{232}Th	100	$(6\alpha - 4\beta)$ series decay	^{208}Pb	$4\cdot92 \times 10^{-11}$	$1\cdot41 \times 10^{10}$
^{87}Rb	27·85	β emission	^{87}Sr	$1\cdot39 \times 10^{-11}$	$5\cdot0 \times 10^{10}$
^{40}K	0·012	11% electron capture	^{40}Ar	$0\cdot585 \times 10^{-10}$	$1\cdot31 \times 10^{9}$
		89% β emission	^{40}Ca	$4\cdot72 \times 10^{-10}$	

basis of the radiometric dating methods.

When a parent isotope disintegrates, very often the initial daughter product is also radioactive, and it decays, perhaps through a series of radioactive nuclei, until the stable end product is reached. A state of equilibrium can be said to exist between the parent element and the daughter products when as many atoms of a member in the series are being formed as are disintegrated per second. This fact is especially important when prospecting for uranium. From observed radioactive intensity an estimate of the uranium content cannot be reliably made unless the degree of radioactive equilibrium in the deposit can be ascertained.

The emission of α, β, and γ-rays associated with the spontaneous disintegration of naturally occurring radioactive substances is of direct importance in the search for radioactive ore bodies. Radioactive prospecting is based on the physical detection of this radiation. Since α and β particles are readily stopped, even by the thinnest cover of overburden, it is mostly through the detection of γ radiation that radioactive deposits can be located.

The standard unit for measuring γ radiation is the röntgen (R). It is the quantity of radiation which will produce $2 \cdot 083 \times 10^{15}$ pairs of ions per m^3 of air at normal temperature and pressure (NTP). In geophysical work, a smaller unit, micro-röntgen ($1 \ \mu R = 10^{-6} \ R$) is more commonly used.

RADIOACTIVITY OF ROCKS

The distribution and intensity of the radioactivity of rocks are subjects of great importance to geophysicists, particularly in the selection of rock material for absolute geological dating and in the calculation of the heat production within the earth (see p.328). They also concern the exploration geophysicist in his search for deposits of radioactive minerals.

Measurements of the radioactive properties of naturally occurring substances indicate that a low level of activity is present in almost all rocks and minerals. In the beginning this activity was attributed almost entirely to traces of uranium and thorium and their radioactive decay products. Later investigations showed that an isotope of potassium (^{40}K) is also radioactive. Although this isotope forms only about 0·012% of the potassium in the earth's crust, it contributes very significantly to the radioactivity of rocks because of the very widespread occurrence of potassium itself in the crustal rocks.

TABLE IX

Typical abundances of important radioactive elements in various rock types (data from Moxham, 1963, and Sass, 1971)

Rock type	U (ppm)	Th (ppm)	K (%)
Granite	5	18	3·8
Sandstone	0·5	2	0·6
Shales	4	12	2·7
Limestone	1·3	1	0·3
Beach sands	3	6	0·3
Basalt	0·5	2	0·5
Dunite	0·005	0·02	0·001
Eclogite	0·04	0·15	0·1

Table IX gives an idea of the average uranium, thorium and potassium content of common rock types. Note that granites and shales show the largest radioactivity. Also, there is a distinct difference in the radioactivity between basalts and granites. This fact is of great significance in the interpretation of terrestrial heat flow data (see p.328).

HEATING DUE TO RADIOACTIVITY

It has been mentioned previously that the radioactive elements emit fast α and β particles and γ-rays, all of which are absorbed by the surrounding medium with the production of heat. The radioactive isotopes, which contribute significantly to the present heat production within the earth are, ^{238}U, ^{235}U, ^{232}Th and ^{40}K. They have half-lives comparable to the age of the earth and hence they are still sufficiently abundant to be important heat sources.

In principle it is possible to measure the heat produced by a given mass of radioactive substance in a known time interval. However, it is more accurate to calculate the heat generation from the energies of the particles and radiation for the individual emissions. In the case of the uranium–thorium series, 90% of the energy is provided by the α particles; the β particles and γ-rays make small contributions. In the case of the decay of ^{40}K, energy is carried both by β particles and by a γ-ray emission associated with a K-electron capture process. An excellent discussion of heating due to radioactive elements can be found in Birch (1954). His values for the heat production by the long-lived isotopes are shown in Table X. In this table, the heat production of ordinary uranium and potassium is based on the known abundance ratios of the isotopes of these elements. The heat production of a given rock type (e.g. granite and basalt) depends on the concentration of the radioactive elements in the rock (see Table IX). From the data of Tables IX and X, the heat production present in rocks can be estimated. This point is discussed further in Chapter 7.

RADIOMETRIC DATING METHODS

Most radioactive age determinations depend upon the production, at a known rate, of a daughter isotope from its radioactive parent since the mineral was crystallized. Experimentally determined decay constants, λ, are available for nuclei used for radiometric dating (see Table VIII). Mass spectrometry permits determination of atomic proportions of parent, P, and daughter nuclei, D, in a mineral or rock sample. The age

TABLE X

Rates of heat production by radioactive elements (after Birch, 1954)

Element or isotope	Abundance (%)	Heat production[1] (cal/g yr)	
^{238}U	99·27	0·71	(0·69)[2]
^{235}U	0·72	4·3	(4·34)
U	natural	0·73	(0·72)
Th	100 (^{232}Th)	0·20	(0·19)
^{40}K	0·012	0·22	
K	natural	27×10^{-6}	(26×10^{-6})

[1] To convert to SI-units (W/kg) multiply by $1·33 \times 10^{-4}$.
[2] Revised values based on the latest decay schemes (quoted in Rybach, 1973).

can then be determined from the following relationship:

$$t = \frac{1}{\lambda} \ln \left(1 + \frac{D}{P} \right) \qquad (6.4)$$

where t is the radiometric age in years, of an event recorded in the sample.

The types of the geological events which can be dated by these methods are: (1) the crystallization of igneous rocks from a magma, (2) the recrystallization of pre-existing rocks, (3) uplift, cooling, and erosion of mountain chains, and (4) the deposition of a sedimentary rock, provided that a new mineral is formed during, or very soon after, sedimentation.

In considering age determinations it is important to emphasize that a good result depends first of all on reliable chemical and isotopic analyses of an *unaltered* mineral specimen, but this is far from the sole criterion. If the age determined is to be useful, the exact location of the sample must be known, as well as particulars about its petrological origin and geological relationship to its surroundings. The choice of samples can best be carried out with the collaboration of a geologist, who can then aid in the proper interpretation of the results obtained.

A comprehensive account of the various radiometric methods applied in geochronology will be found in York and Farquhar (1972). We shall briefly discuss here only some important points regarding the applicability and limitations of the most commonly used methods.

Potassium–argon method

Potassium is a common and widespread element in the crustal rocks. The isotope ^{40}K ($0\cdot\dot{1}19\%$ of natural K) decays in two ways (see Table VIII), of which only the ^{40}K–^{40}Ar decay process is used for dating. The decay to ^{40}Ca is not used for dating, because non-radiogenic ^{40}Ca is generally initially present in potassium-bearing minerals. Since argon is not a common constituent of minerals, errors due to the presence of initial ^{40}Ar are usually (but not always) small. Correction for the contamination by atmospheric argon may be necessary and can be made by applying the present atmospheric ratio, ^{40}Ar/^{36}Ar $= 295\cdot5$.

Argon produced by the decay of ^{40}K tends to diffuse from the host; above 300°C the rate of diffusion is appreciable in most minerals. Commonly occurring igneous or metamorphic minerals, which are suitable for K–Ar dating, are micas and hornblende. High-temperature forms of the potassium-feldspars (e.g. sanidine) and plagioclase feldspars are generally suitable. In contrast, the most common potassium-feldspars, such as orthoclase and microcline, are unsuitable because they can lose argon readily even at atmospheric temperatures. Whole-rock analyses of rocks, such as basalts, yield reliable dates only if the rock has suffered no alteration and is free of glass in the matrix. The dated event is the time of final cooling below approximately 200°C. The marine sediments, limestone, and sandstone, sometimes contain the potassium-bearing mineral glauconite, which is formed at the time of sedimentation and can be dated by the K–Ar method.

The great advantages of the K–Ar method are the abundance of potassium and the relatively short half-life of ^{40}K. The method can be used to date a wide variety of rocks, of almost all geological ages, ranging from the oldest terrestrial rocks (ca. 3700 m.y. old) to the most recent rocks which are sometimes as young as 30,000 years. K–Ar dating of younger rocks has been of great advantage, particularly in establishing the chronology of the recent reversals of the earth's magnetic field (see Fig.124, p.240). No other method is available for such young rocks, and there is in fact a gap in the availability of absolute ages between the age of youngest rocks dateable by the K–Ar method and the age of the oldest material which can be dated by the ^{14}C method.

Rubidium–strontium method

Although rubidium in natural abundance cannot compare with potassium, it is widely distributed among minerals such as the micas and feldspars. The decay scheme of ^{87}Rb (see Table VIII) is a simple one consisting of the emission of a weak β particle to form ^{87}Sr.

There is still no agreement on the exact value of the half-life of ^{87}Rb because of the difficulties to determine this by direct radioactivity measurements. The two values in current use are 47,000 m.y. and 50,000 m.y.

The Rb–Sr method can be used to date such common rock-forming minerals as micas, and all types of potassium-feldspars, including orthoclase and microcline. For whole-rock analyses, granites and granitic gneisses are particularly suitable.

The great advantage of the Rb–Sr method is that it represents a solid-solid system and as such there is relatively less chance that the parent or daughter product will be lost. The disadvantages are that Rb is not an abundant element in the crust, and the long half-life makes it difficult to apply to young rocks; also the initial presence of non-radiogenic Sr in most minerals requires correction. It is, however, probably the most suitable method for dating shield areas of Precambrian rocks.

Uranium–lead method and the age of the earth

Uranium and thorium frequently occur together in the same mineral and their isotopes (^{238}U, ^{235}U, and ^{232}Th) decay (see Table VIII) to form lead (^{206}Pb, ^{207}Pb, and ^{208}Pb). The formation of this lead, called radiogenic lead, is the basis of the uranium methods of age determination.* Since the radioactive isotopes, with different decay constants, are available in any uranium, a check on the consistency of the deduced ages is possible. This is not the case in the Rb–Sr or K–Ar methods.

The mineral most commonly used for U–Th–Pb dating is zircon ($ZrSiO_4$). Zircon occurs as an accessory mineral in many rocks and contains a small fraction of uranium atoms substituted for zirconium. Lead resulting from the uranium decay is trapped in the crystal and can be preserved through metamorphic events.

Besides the radiogenic isotopes (^{206}Pb, ^{207}Pb, and ^{208}Pb), lead has another isotope, ^{204}Pb, which is non-radiogenic. The latter is not a product of radioactive disintegration. By consideration of the isotope ratios of lead itself (i.e. ^{206}Pb/^{204}Pb, ^{207}Pb/^{204}Pb, and ^{208}Pb/^{204}Pb) it is possible to deduce an age for the earth, and also to trace the history of a particular lead sample. The method is fairly complex so that only a brief outline can be given here.

In the primeval earth, naturally occurring lead must have possessed certain quantities of ^{204}Pb, ^{206}Pb, ^{207}Pb, and ^{208}Pb. With the passage of geological time, the quantity of ^{204}Pb (non-radiogenic) must have remained unchanged while that of the other three must have increased because of the disintegra-

---

*Another uranium method of dating is based on the spontaneous fission of ^{238}U and neutron-induced fission of ^{235}U. When the product particles of fission travel through a material, they leave records of their tracks whose density is a measure of the concentration of uranium. For more details of the fission-track dating method see York and Farquhar (1972).

tion of uranium and thorium. Actual comparison of the isotopic compositions of lead minerals (such as galena) of different
ages revealed that ^{206}Pb, ^{207}Pb, and ^{208}Pb have indeed increased
in quantity with time (Fig.157). Once the rate of increase is

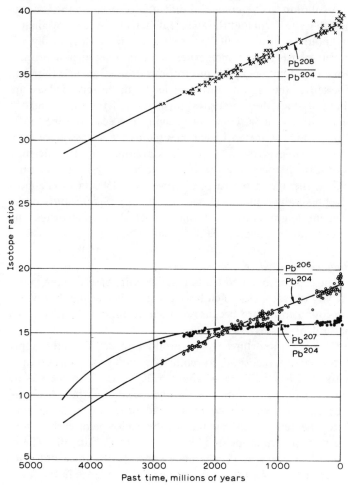

Fig.157.Variation in the isotopic composition of lead with age. (After
Jacobs et al. 1959.)

quantitatively determined, the age of the earth can be extrapolated. In actual practice, this gave uncertain results because the initial quantities of ^{206}Pb, ^{207}Pb, and ^{208}Pb, existing at the time of the earth's birth, were unknown.

A solution to overcome this difficulty is based on the assumption that the earth and the ancestral meteorite-planet originated at about the same time. Certain meteorites, specifically the triolite-phase of iron meteorites, contain a significant proportion of lead but negligible proportions of uranium and thorium. Therefore, the isotope ratios of lead can be regarded as unchanged since the meteorites' formation. If the earth and the meteorites had a common origin, the lead-isotope ratio thus fixed in a meteorite can be used as the lead-isotope ratio in the primeval earth. With the development of sufficiently precise techniques for the lead-isotope analysis in the early 1950's, the age of the earth obtained was 4500 m.y. The most accepted value today is closer to 4600 m.y. Significantly, independent determinations of the age of stony meteorites and several lunar samples yielded similar values.

Radiocarbon and tritium methods

The radiocarbon and tritium methods are based on the radioactivity of isotopes, which have a much shorter life, but which are found in nature because of their continual production by natural processes.

Of great importance is the production of ^{14}C in the upper atmosphere by cosmic-ray collisions with ^{14}N. ^{14}C is radioactive, with a half-life of about 5730 years. It mixes readily with the ordinary atmospheric and oceanic carbon (^{12}C and ^{13}C) and is taken up by plants and other living organisms. After the death of the organisms, the exchange of carbon with the atmosphere stops and the ^{14}C supply is cut off. Thereafter ^{14}C decays at its characteristic rate. Radiocarbon ages are calculated from the ratio of the ^{14}C abundance in the fossil specimen to the ^{14}C abundance in similar living material:

$$\frac{^{14}C \text{ (fossil material)}}{^{14}C \text{ (living material)}} = e^{-\lambda t} \qquad (6.5)$$

Samples used for radiocarbon dating must satisfy the condition that the ^{14}C now present is the same carbon that was present in the material at the time of its death. Libby (1955) discussed the assumptions involved and the techniques of measurement. Because of its short life, ^{14}C obviously cannot be used for dating much beyond 30,000 years. This restricts the use of the method for the study of extremely recent geological events. The ^{14}C method has mainly been used for dating plant material. Apart from this, of particular importance is the dating of the mollusc shells in postglacial beach deposits, which has been used to determine the rate of vertical movement of the land surface (e.g. Fennoscandia, see p.123). The method has also been used with success for dating glacier ice from icebergs in Greenland (Scholander et al., 1961). Glacier ice contains bubbles of atmospheric air trapped at the time of formation of the ice. ^{14}C dating of the CO_2 content of the bubbles thus gives the age of the ice.

Tritium, a radioisotope of hydrogen (^3H), is also produced in the upper atmosphere as a result of the cosmic-ray activity. The half-life of tritium is about 12·5 years; even in this short time tritium becomes rapidly mixed with the active hydrogen reservoir of the earth. Circulating waters maintain their tritium activity by the continual addition and mixing of rain. In isolated water, the tritium content decays exponentially. The method opens the possibility of studying the water cycle over the continents and the oceans and of determining the characteristics of underground waters with respect to their "age" and "recharge" rates. This potentiality was first realized by W.F. Libby in 1951 who made extensive measurements of the ^3H contents of a variety of waters. Unfortunately, tritium is a product of thermonuclear explosions, and that has severely limited its usefulness in geologic studies. The only regions, where natural pre-bomb levels of ³H may still be found today, are the deep-seated old waters on continents or in deep oceans which are unaffected by human activity due to the migration of water.

GEOLOGICAL TIME SCALE

Before the advent of radiometric dating methods, the ages of geological formations in different parts of the world were correlated using index fossils preserved in the sedimentary strata. The fossil record, with the exception of a very limited range of microfossils, extends only as far back as the early Cambrian period, covering the Phanerozoic eon of about 600 m.y.

From the time absolute ages first became available, an attempt has been made to relate these ages to the stratigraphical time scale. Unfortunately few sedimentary rocks and few minerals occurring in them can be directly dated by radiometric methods. Most of the best age determinations have been made on crystals obtained from pegmatites occurring around the margins of large intrusive bodies. In many cases a pegmatite or igneous rock intersects older sedimentary formations and is overlain by younger sedimentary formations, and thus an age is obtained which lies somewhere in the gap between the two formations of sedimentary succession.

Holmes (1947) established the first absolute time scale based on radiometric dates. Kulp (1961) reviewed the reliable ages that were available for rocks whose stratigraphic positions were known, and prepared another time scale. Table XI shows the scales of Kulp (1961) and Holmes (1965) together with the time scale which is the most widely used at present (Harland et al., 1964). The table also includes the age of the earth, and the age of the oldest rock dated so far (Moorbath et al., 1972).

PRECAMBRIAN CHRONOLOGY

One cannot fail to notice the great duration (ca. 4000 m.y.)

TABLE XI

The geological time scale with absolute ages in millions of years before present*

Geological Period	Kulp (1961)	Holmes (1965)	Harland et al. (1964)
Cenozoic (mammals)			
Quaternary			
Recent			
Pleistocene	1	2 to 3	1·5 to 2
Tertiary			
Pliocene	13	12 ± 1	~7
Miocene	25	25 ± 2	26
Oligocene	36	40 ± 2	37 to 38
Eocene	58	60 ± 2	53 to 54
Palaeocene	63	70 ± 2	65
Mesozoic (Reptiles)			
Cretaceous	135	135 ± 5	136
Jurassic	181	180 ± 5	190 to 195
Triassic	230	225 ± 5	225
Palaeozoic (Invertebrates)			
Permian	280	270 ± 5	280
Carboniferous	345	350 ± 10	345
Devonian	405	400 ± 10	395
Silurian	425	440 ± 10	430 to 440
Ordovician	500	500 ± 15	~500
Cambrian	600 ?	600 ± 20	570
Precambrian (no developed fossils)			
Oldest dated rock	3700 m.y. B.P.		
Age of the earth	4600 m.y. B.P.		

*Ages correspond to the lower boundaries of geological periods.

of the Precambrian that represents 87% of what we now understand to be the age of the earth. Certainly one of the great contributions of radiometric chronology has been the dating of events in Precambrian rocks older than 600 m.y. Although divisions of the Precambrian do not yet have a universal nomenclature, age determinations formed the basis of dividing the Precambrian shield areas of the continents into provinces, each characterized by tectonic activity.

Dearnley (1966) showed that if a histogram is made of all available age determinations, three Precambrian peaks occur (Fig.158) corresponding to the tectonic activity beginning about 2700 m.y., 1900 m.y., and 1000 m.y. ago. Significantly, in every continent, except Antarctica, rocks are known to have survived a great length of time (2000 m.y.) without later metamorphism. This has led to the speculation that these small areas with the oldest rocks are the nuclei around which the continents have grown with time.

Not all geologists have accepted the idea of continental growth by lateral accretion. Some see the Grenville province

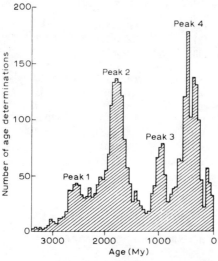

Fig.158. Frequency histograms of igneous and metamorphic age determinations. Three Precambrian peaks occur, corresponding to tectonic activity beginning about 2700, 1900, and 1000 m.y. ago. (After Dearnley, 1966; reproduced from Bott, 1971.)

in the Canadian shield (age $\sim$ 900–1100 m.y.) as an ancient basement, partially covered with late Precambrian sediments, which was completely metamorphosed and reworked during the 1000-m.y. orogenic event. Also, the very old ages, sometimes observed on the periphery of continents (e.g. the San Gabriel Mountains of California with ages of about 1500 m.y.), do not substantiate the continental-growth hypothesis.

The issue of continental growth around a nucleus is far from settled, but it can hardly be contested that radiometric dating has established three major tectonic events during the earth's geological history in the Precambrian.

OXYGEN ISOTOPES AND PALAEOCLIMATES

The study of palaeoclimatology is deeply rooted in geology, and in the recent decades there has been an impressive accumulation of data on various palaeoclimatic indicators (Nairn, 1964; Bowen, 1966). The most direct quantitative estimates of ancient climatic features are the palaeotemperatures derived from analyses of oxygen isotopes in the fossil shells of certain marine organisms. The $^{18}O/^{16}O$ ratio in shells of pelagic and benthonic foraminifers depends on the temperature of the formation and on the isotopic composition of the ambient water. The natural $^{18}O/^{16}O$ ratio is approximately 1/500. Measurement of the ratio to one part in 10,000, corresponding to temperature variations of $\pm0.5°C$, demands very sophisticated mass spectrometry techniques.

The first detailed studies on deep-sea cores from the Pacific and Atlantic Oceans and Carribean Sea were made by Emiliani (1955). These are tropical regions where the annual temperature variation is relatively slight, so that no uncertainty arises from the selective seasonal growth of the carbonate deposits. Temperature variations over the past 300,000 years, as obtained by Emiliani, are shown in Fig.159. The results indicate that the climatic fluctuations responsible for the Pleistocene glaciation were worldwide phenomena.

312

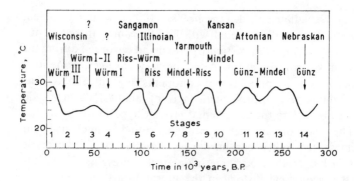

Fig.159.Tropical, deep-ocean temperature variations over the past 300,000 years, obtained from the ratios of ^{18}O and ^{16}O isotopes. (After Emiliani, 1955.)

The ^{18}O concentration in glacier ice has also been used for studying palaeotemperature variations (Dansgaard et al., 1969). In Fig. 160 the relative temperature curve, based on the ^{18}O analysis of the Greenland ice core and the foraminiferal climate curve in the deep-sea sedimentary core from near the Cape Verde Islands, are correlated with the atmospheric radiocarbon activity and the earth's magnetic intensity variations. A climatic optimum is indicated at about 6500 B.P. The frequent oscillations in the ^{18}O temperature curve are probably related to fluctuations in the solar radiation which also cause corresponding variations in the ^{14}C concentration of the atmosphere. Note the striking correlation between the warmer climate, the increased radiocarbon activity and the decreased geomagnetic field intensity around 6500 B.P.

RADIOACTIVITY SURVEYING

The prospecting methods employing radioactivity have become important in recent decades, not only for meeting the growing demand for uranium, but also for the location of associated minerals and ore bodies. The methods can often also

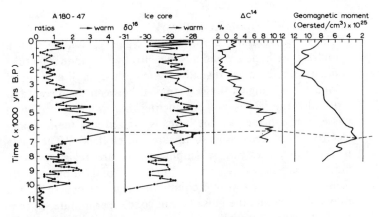

Fig.160. Correlation of the foraminiferal climate curve in deep-sea core A 180-47, the oxygen-isotope climate curve in an ice core from Greenland, the variation in atmospheric radiocarbon activity, and the changes in magnetic intensity of the earth. (After Wollin et al., 1971.)

be used with advantage in geological and structural investigations.

It has been mentioned previously that the geophysical search for radioactive elements in the earth's crust primarily involves the locations of areas with abnormally high γ radiation. This is because α and β-rays are comparatively much less penetrating, and they will be indetectable as soon as a radioactive deposit has an overburden cover of a few centimetres. For detailed information on the various aspects of radioactivity prospecting, the reader is referred to Jakosky (1950), Dobrin (1960), and Lang (1970). Techniques of the subject are reviewed in the 1969 volume of the International Atomic Energy Agency (I.A.E.A.).

The following is a very brief and elementary discussion of the subject and omits important details.

Instruments and field procedure

Of the many types of radiation detectors designed for field use, the Geiger counter and the scintillation counter are the most suitable for prospecting work.

The Geiger counter consists of a sealed glass tube with a cylindrical cathode around a thin central wire anode. The tube is filled with some gas (usually argon with a small amount of alcohol) and a high voltage is applied between the electrodes. Normally the gas is non-conducting but when γ radiation passes through the gas, the latter is ionized and the ions and electrons produced accelerate towards the electrodes. The resulting current pulses can be amplified and are registered on a meter or are heard as a "click" in a pair of headphones. By using a suitable electronic circuit the instrument works as a ratemeter, reading the pulse rate in counts per minute. In contrast to α and β particles, the γ-rays are very weakly ionizing, and, therefore, the γ-ray detection efficiency of a Geiger counter is very low.

The scintillation counter is a more efficient type of detector. It utilizes the fact that certain crystals such as thallium-activated sodium iodide emit a visible flash of light (scintillate) when they absorb γ-rays. The scintillations can be detected by photomultiplier tubes, and after suitable amplification can be read on a meter in counts per minute.

Because of its much greater detection efficiency (almost 100% for detecting γ radiation), the scintillation counter has in many cases replaced the Geiger counter in modern work. Scintillation counters are nowadays used almost exclusively in airborne surveys to measure the surface radioactivity from the air. More recently scintillation detectors have been designed to differentiate the counts due to γ radiations of different energy and this, for example, enables one to distinguish between uranium and thorium deposits.

Ground-radioactivity surveys may be carried out by walking along lines holding the detector about half a metre above the ground. If there is evidence of a radioactive occurrence in an area, detailed measurements can be made in a grid pattern at close intervals of a few metres. If the detector records a count rate several times greater than the "background", a radioactive deposit may be indicated. The background effect is mainly due to cosmic radiation and potassium which is relatively abundant in granitic rocks. An outcropping pegmatite dyke with a high potassium-feldspar content might give the same count rate as a uranium deposit less than one-third metre below the ground, and using only a scintillation counter it is not possible to distinguish between the two. Measurements with a γ-ray spectrometer make it possible to distinguish between the activities of different sources and thereby enable in-situ assaying of uranium and thorium. Fig.161 shows a typical field setup of a portable γ-ray spectrometer.

Fig.161. A portable gamma spectrometer in use in Kvanefjeld, South Greenland. The instrument has a collimated detector unit providing a constant solid angle of detection in a rough terrain. (Courtesy of L. Løvborg.)

Airborne-radioactivity surveys are often carried out in conjunction with airborne magnetic and electromagnetic surveys for mineral prospecting. A number of large scintillation sensors, coupled in suitable arrangement, are employed to enhance the detection sensitivity of the counter unit, and the count is recorded automatically on a moving strip of paper. Flight altitude above the ground surface is usually not greater than 100 m and preferably less if very small uraniferous outcrops are sought. A very instructive article on airborne radiometric surveying is given in Pemberton (1970).

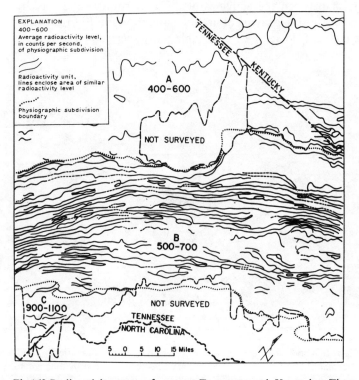

Fig.162. Radioactivity map of eastern Tennessee and Kentucky. Three radioactivity provinces, as distinguished by the average radioactivity level and a distinct pattern, outline three physiographic subdivisions: *A* = the Cumberland Plateau, *B* = the Valley and Ridge province, *C* = the Blue Ridge province. (After Bates, 1966.)

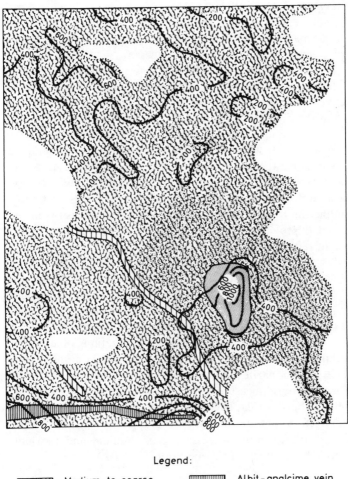

Legend:

Medium-to-coarse grained lujavrite

Preagpaitic inclusion

Albit-analcime vein with steenstrupine

Albit-analcime vein without steenstrupine

—400— Contour line, ppm U

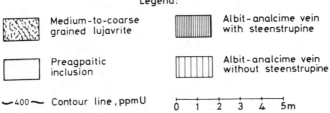

0 1 2 3 4 5m

Fig.163. Contour lines of the uranium concentration (in ppm) superimposed on the geological map of a mineralized area in the Ilímaussaq intrusion, South Greenland. (After Løvborg, 1972.)

Examples of radioactivity surveys

Radioactivity surveys are an extremely useful tool in geological mapping and exploration programmes. Two examples are given to illustrate their usefulness.

The first example is from the Tennessee–Kentucky area and is taken from a paper by Bates (1966). Fig.162 is a radioactivity map of the area. The most obvious feature of the map is the three areas (*A*, *B*, and *C*) with different patterns and levels of radioactivity. These three radioactivity provinces closely coincide with the boundaries of the three physiographic subdivisions within the area. Area *B* has an overall moderate radioactivity level with a strong linear pattern of radioactivity. In several cases linear high-level radioactivity belts accurately outline the traces of thrust faults. The boundary between the *B* and *C* provinces (Fig.162) is fairly well defined by a sharp rise in radioactivity level of the Blue Ridge strata.

The other example is from the Kvanefjeld plateau in the Ilímaussaq alkaline intrusion in South Greenland. As described by Sørensen et al. (1969), the highest concentrations of uranium and thorium occur in the lujavrites in the northernmost region of the intrusion, where the lujavrites are in direct contact with the roof rocks. The most intense mineralizations of uranium and thorium are associated with steenstrupine, a uranium-rich variety of monazite and thorite. Fig.163 shows the results of 1-m-spaced grid measurements with a portable γ-ray spectrometer (Løvborg, 1972). There is a good correlation between the radiometric contours and the lithology. The example shows the advantage of in-situ assaying of uranium and thorium, especially in difficult field conditions.

CHAPTER 7

Geothermal Methods

INTRODUCTION

The heat felt at the earth's surface comes mainly from the sun. However, most of the solar heat is radiated back into space and only a very minute fraction is able to penetrate below a depth of a few hundred metres. Thus its influence on the interior of the earth is negligibly small in comparison with that of the heat occurring within the earth.

The main source of heat energy within the earth is presently believed to be the radioactive decay of long-lived isotopes, but other sources of heat, such as the initial temperature and the heat released by accretion (due to the gravitational work performed in assembling and compacting a body from small constituents), may have been substantial early in the earth's history.

Heat is gradually and sometimes dramatically transferred from the earth's interior to the surface. The most dramatic examples are volcanoes and hot springs. The heat escaping through the earth, directly or indirectly, sets off various geological processes, which are related to tectonic movements and igneous and metamorphic activity.

The study of thermal behaviour of the earth is one of the most speculative branches of geophysics. Below a depth of about 100 km, the temperature distribution is highly uncertain and the distribution of heat sources and the mechanisms of heat transfer are unknown. Nevertheless, the subject is of great interest because of its important bearing on all the hypotheses concerning the origin and development of the earth.

In recent decades extensive studies of the heat outflow from the earth have provided information on the broad regional characteristics of the thermal conditions beneath the major geological features of continents and oceans. This information is of special significance to the theory of convection in the mantle with which the hypotheses of continental drift, sea-floor spreading and plate tectonics are closely interlinked. On a local scale, thermal measurements (in particular temperature anomalies) have been used to infer the position of structures such as shallow salt domes, anticlines, faults, fissures, etc. In addition, temperature measurements in boreholes have been used to correlate stratigraphic horizons.

This chapter aims to provide a brief and elementary discussion of the geothermal methods and their applications to both global and local geological problems.

THERMAL PROPERTIES OF ROCKS

As has been observed in boreholes and mines almost everywhere on the earth, the temperature increases with depth. Therefore, heat must be flowing upward in the earth. The amount of heat flow, q, depends on the thermal conductivity, K, of the rock material and the temperature gradient involved. The working relationship is:

$$q = -K \operatorname{grad} T \qquad (7.1)$$

where grad T $(= \partial T/\partial Z)$ is the rate of increase of the temperature with depth. q is measured in units of energy per unit area per unit time, the SI-unit being W/m^2. In the c.g.s. system, the working unit is micro-calories per cm^2 per second ($\mu cal/cm^2$ s).

The thermal conductivity is a measure of how easily heat flows through a material. Some simple observations indicate that the thermal conductivity of the earth is low; for example, the daily variations of the surface temperature are hardly felt at a depth of 1 m and the effects which do penetrate arrive half a day or more late. A few metres further below the surface only the seasonal changes in temperature can be detected, and they arrive months late. At a depth of approximately 1 km evidence of a low temperature remaining from the last glacial age (about 10,000 years ago) can still be found.

The lingering effect of temperature with time depends on the thermal diffusivity, α. It is defined as:

$$\alpha = K/\rho c_p \tag{7.2}$$

where K is the thermal conductivity, ρ the density and c_p the specific heat at constant pressure. The SI-unit for α is m^2/s.

The thermal diffusivity (as well as the thermal conductivity) is very low for most rocks, the range for α being $0.5-2 \times 10^{-6}$ m^2/s, or, on a geological scale, $15-60$ $km^2/m.y.$ This means that a thermal event originating at a depth of about 100 km will not be perceptible near the surface for somewhere between 10 and 100 m.y., if the heat were to be transferred by conduction alone.

Table XII gives typical values for the thermal conductivity of different rocks and minerals. Conductivity is to a great extent controlled by the minerals present in a rock, but the effects of the porosity and the increase in temperature and pressure can be significant. A good estimate of thermal conductivity is 2.5 W/m °C down to about 50 km depth; below this depth the thermal conductivity is uncertain. More data on the thermal conductivities of rocks and minerals can be found in Clark (1966).

TEMPERATURES WITHIN THE EARTH

The simplest way of studying the earth's temperature is to drill a hole and to use a sensitive thermometer or a thermistor probe. It is not always necessary to drill, since existing mine

TABLE XII

Thermal conductivities of rocks and minerals at normal temperature and
pressure (data from Clark, 1966, and Parasnis, 1971)

Material	Conductivity in SI units* (W m °C)	
Granite	1·9–3·2	(2·7)
Granodiorite	2·6–3·5	(3·0)
Gneiss		
// to foliation	2·5–3·7	(3·1)
⊥ to foliation	1·9–3·2	(2·7)
Basalt	1·5–2·2	
Diabase	2·1–2·3	(2·2)
Gabbro	2·0–2·3	(2·15)
Serpentinite	2·0–3·8	(2·3)
Dunite	3·7–5·2	
Sandstone	2·5–3·2	
Shales	1·3–1·8	(1.4)
Limestone	2·0–3·0	(2.5)
Rock salt	5·3–7·2	(5.7)
Haematite ore	10·5	
Haematite crystal		
// c	12·1	
⊥ c	14·8	
Magnetite (polycrystalline)	5·3	
Water	0·59	(25°C)
Ice	2·2	(0°C)

*To convert to c.g.s. units (10⁻³ cal/cm s °C) multiply by 2·39.

shafts, tunnels, and oil wells can be utilized for this purpose.
Measurements obtained in this manner show that the earth's
temperature at any locality increases with depth, the average
rate of increase (the so-called geothermal gradient) being
about 3 C per 100 m of depth in non-volcanic areas.

A borehole is at most a few kilometres deep. How can we estimate the earth's temperature beyond that depth? The surface heat flow provides a clue for deducing the temperature in the crust. If we assume that the geothermal gradient continues at the same rate ($3°C/100$ m) all the way to the bottom of the crust (33 km deep), the temperature there would be around $1000°C$. However, the thermal gradient at the bottom of the crust must be smaller than the surface gradient, the difference being due to the radioactive heat generated in the crust. This implies a decrease of thermal gradient with depth. To some extent this decrease is offset by the fact that the thermal conductivity tends to decrease as temperature rises; therefore, the greater the depth, the greater would be the thermal gradient to convey the same quantity of heat. Taking both factors into consideration, the temperature at the base of the continental crust is estimated at $600° \sim 800°C$. At sea, since the crust is only about 6 km thick, the temperature at its base is thought to be about $150° \sim 200°C$. Various models of temperature distribution below the continents and oceans, along with the assumptions involved, have been discussed by Macdonald (1965).

To estimate the temperature in the mantle and the core we have to resort to indirect methods. Inferences about the temperature distributions at greater depths are based on the observed seismic velocities and the variations in electrical conductivity, but these depend on the assumptions of the physical properties of the postulated mantle material (peridotite?) at higher pressures and temperatures. Nonetheless, these characteristics can set the upper and lower limits of the temperature estimates.

Seismological data suggest that the mantle is essentially crystalline and that the outer core is liquid. The temperature within the mantle is, therefore, below the melting-point curve for mantle. For the outer core, the melting point of iron under pressure prevailing in the core would give the minimum temperature. Likewise the melting point for iron corresponding to the existing pressure in the inner core would give the maximum temperature for the solid inner core.

Estimates of the earth's temperature obtained by two methods are shown in Fig.164. Within the margins of uncertainties involved, the hatched curve should serve to give a general picture of the temperature distribution inside the earth. For an excellent discussion of the subject the reader is referred to papers by Gilvarry (1957) and Verhoogen (1960).

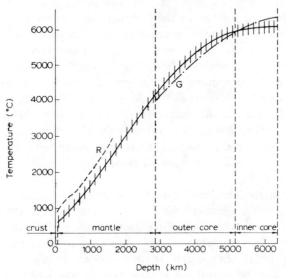

Fig.164. Estimates of the temperature in the earth's interior. Curve *R* (after Rikitake, 1952) is based on the variations of electrical conductivity in the mantle. Curve *G* (after Gilvarry, 1957) for the core is based on estimated values of the melting point of iron.

TERRESTRIAL HEAT FLOW

The heat that flows from the earth's interior to its surface and escapes into space is called the "terrestrial heat flow". This is the quantity of thermal energy that the earth is losing, in other words, the expenditure of the earth's heat budget. The present rate of heat loss from the earth's surface is about 1×10^{21} joules/year*. In terms of energy this outflow of heat is the most impressive of geophysical processes; the energy loss involved in earthquakes and volcanic activity is orders of magnitude lower.

Heat flow measurements

A determination of heat flow requires two separate measurements: the thermal gradient, $\partial T/\partial Z$, and the thermal conductivity, K, of the rocks in which the temperatures are measured. Heat flow across a unit area is then calculated by the formula $q = K \partial T/\partial Z$.

A probe device for measuring the heat flow in soft sediments on the ocean floor is illustrated schematically in Fig.165. After the probe is inserted, the temperature gradient in the sediments is found from the temperature measurements along the probe. Then electric heating is applied to the bottom of the probe at a known rate, and after a steady state is reached, the temperature gradient is again measured. The thermal conductivity, K, is calculated from the second set of observations. On land, mines and boreholes are used to find the temperature gradient, and direct laboratory determinations are made of the thermal conductivity of collected rock samples.

Measurements of heat flow commenced on land in the 1930's and up to a short time ago a few hundred measurements had been reported. Although measurements at the sea floor began in 1952 with the pioneer work of E.C. Bullard and his associates, the data obtained at sea now outnumbers those taken on land. This is due to the fact that measurements are much simpler to take at sea than on land; at the sea bottom the temperature of water is usually stable during the entire year, and, therefore, there is no need to drill deep holes.

The measurements confirm that the heat flow is indeed small,

*When expressed per second it is equivalent to $3 \cdot 2 \times 10^{13}$ W.

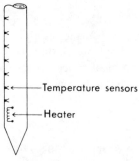

Fig.165.Probe device used to make heat-flow measurements in the soft sediments of the ocean floor.

averaging about 61 mW/m² both for continents and oceans. Thus, the geothermal heat is very small compared to the energy arriving from the sun, too small to have any direct effect on the climate. The observations also show a characteristic variation in the heat flow of different geologic provinces as shown in Table XIII. On the continents, Precambrian shields (tectonically stable regions) have low heat-flow values. The orogenic areas, by contrast, show higher heat-flow values; the more recent the volcanic origin or orogeny of an area, the higher its heat-flow is likely to be. In the oceans, ridges are provinces of a high but variable heat flow, basins are characterized by a moderate and uniform heat flow and trenches are low heat-flow provinces. For a detailed discussion of the global heat-flow data the reader is referred to the A.G.U. Monograph edited by Lee (1965).

Equality of continental and oceanic heat flow

The most important and most unexpected information from the data obtained (Table XIII) is that the mean heat-flow values of continents and oceans are nearly equal. This came as a great surprise to the geophysicists.

Prior to the actual measurements, it was thought that the oceans, with their thin (about 5 km) "mafic" crust and a lower concentration of radioactive elements, would have a much lower heat flow than the thick (30–35 km) "silicic" continental

TABLE XIII

Mean heat flow values (q) for various tectonic regions of the earth[*1]
(data reproduced from Sass, 1971)

Tectonic region	No. of data	q (mW/m^2)[*2]	Standard deviation
All continents[*3]	597	61·0	19
Precambrian shields	214	41	10
Post-Precambrian non-orogenic areas	96	62	17
Palaeozoic orogenic areas	88	60	17
Mesozoic–Cenozoic orogenic areas	159	74	24
All oceans[*3]	2530	61·4	33
Ocean basins	683	53	22
Mid-oceanic ridges	1065	80	62
Ocean trenches	78	49	29
Continental margins	642	75	39
World[*3]	3127	61·4	31

[*1]The individual data were weighted according to quality, so that a poorly determined result makes a smaller contribution to the mean than a precisely determined value.

[*2]The SI-unit for heat flow is watt/metre^2; however, it is more convenient to use a sub-unit milliwatt/metre^2 (mW/m^2). To convert these values to c.g.s. units (μcal/cm^{2}s) multiply by 23·9 $\times$ 10^{-3}.

[*3]To compensate for the uneven geographical distribution of data within these larger units, means are determined by giving equal weight to the average heat flow in quadrilaterals of equal area (3 $\times$ 10^5 km^2 or 5° $\times$ 5° at the equator).

crust with a considerably higher concentration of radioactive elements.

Virtually all radioactive heat is produced by the isotopes of uranium, thorium, and potassium. From the data on radioactive heat produced by the main rock types (Table XIV) it can be

TABLE XIV

Estimates of heat production in terrestrial materials[*1]

Material	Composition (ppm)[*2]			Present heat production (10^{-12} W/kg)[*3]
	U	Th	K	
Granite	4·75	18·5	37,900	910
Basalt	0·60	2·7	8400	160
Chondrite	0·012	0·04	845	5·2

[*1]The heat production is to be compared with the mean outflow of heat per kilogramme of the earth (5·4 × 10^{-12} W/kg). This is obtained by multiplying the observed surface heat flow (0·061 W/m^2) by factor S/M, where S is the surface area of the earth, M its mass.
[*2]Compositions taken from Macdonald (1965).
[*3]To convert to c.g.s. units (cal/g s) multiply by 2·39 × 10^{-4}.

shown that a 20–25 km layer of granitic crust could generate all of the observed heat at the continental surface. The heat production in the oceanic crust, which is assumed to be about 5 km of basalt, gives a heat flux of only 1·5 m W/m^2, or less than 3% of the observed oceanic heat flow. The near-equality of continental and oceanic heat flow thus indicated a fundamental difference between the mantles underlying these topographic units.

The most obvious implications are that the oceanic mantle has higher temperatures and a higher concentration of radioactive elements than the continental mantle. An explanation put forth by Macdonald (1965) suggests that virtually all radioactive elements in the continental mantle have risen with the granitic differentiates which formed the continents, whereas these elements are still distributed through the oceanic upper mantle down to perhaps 500 km, the average integrated vertical concentration being the same for both the continents and oceans. This idea implies the permanence of continents and oceans, and suggests that the upper mantle underwent chemical differentiation early in the history. This is particularly difficult to reconcile with the growing body of evidence for continental drift and ocean-floor spreading.

The differentiation hypothesis also requires the oceanic mantle to be less dense, both because it is hotter (and therefore thermally expanded) and because it is presumed to contain the lighter constituents which have been lost by the continental mantle. Such a density difference is incompatible with the form of the geoid (see Fig.69), which shows no distinction between continents and oceans. For this reason it is desirable to look for an alternative explanation.

The alternative explanation of the oceanic-continental heat-flow problem, first suggested by Bullard et al. (1956), is based on the transfer of heat in the upper mantle by convection. This hypothesis supposes that most of the oceanic heat flow is carried through the upper mantle by convection. In the upper mantle beneath the continents, convection is assumed to be absent or to carry a much smaller portion of heat flow reaching the surface. The general picture is (Fig.166) that the convection currents rise near the ocean ridges and discharge heat as they flow towards the continents. The exact mechanism of heat consumption is still subject to much speculation. However, the convection hypothesis adjusts more easily to the modern ideas of continental drift and ocean-floor spreading. This very idea of convection in the upper mantle stimulated the study of the ocean floors. With some constraints the idea seems to be compatible with the recent concept of plate tectonics. The subject of convection in the mantle has been discussed in detail by various authors (e.g. Vening Meinesz, 1964; Elsasser, 1966; Knopoff, 1967; Elder, 1968).

Regions of anomalous heat flow

In many regions where sufficient measurements have been

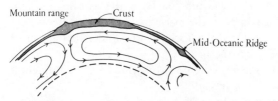

Fig.166. A model of thermal convection in the upper mantle. Convection currents rise near the oceanic ridges and discharge heat as they flow towards the continents.

made, significant variations from the average heat flow have been observed. As might be expected many of the high heat-flow regions are located near the principal ocean ridges. In these regions the heat flow is highly variable and, locally, values as high as five times the global average have been observed. Possibly these regions are equivalent to volcanic or geothermal areas on land, and are attributed to localized sources of heat ("hot spots") situated at shallow depths within a few tens of kilometres from the surface. The magma, which produces the hot spots, is formed at a much greater depth, probably by partial fusion of the mantle material in a rising convection current.

On the continents are many regions of anomalous heat flow, apart from volcanic areas. The most active tectonic regions show the highest and most stable regions (shields) show the lowest values. Young mountain ranges such as the Alps, and also some non-tectonic regions, possess a heat flow which is about 50% above the average for the continents. The high heat flow of young orogenic areas may be mainly a result of crustal thickening during mountain building. Some regions show high heat flow without evidence of crustal thickening. For instance, in southeast Australia the observed heat flow is about 50% above the average continental value (Sass, 1964). This regional anomaly could be caused by an unusually high concentration of radioactive heat sources in the crust or in the upper mantle. Alternatively, it could be attributed to the effect of a convection current in the underlying mantle. The Hungarian basin in Europe provides another example. This is a basin of thick Tertiary sediments surrounded by the Carpathians and the Dinaric Alps. The basin is believed to be of volcano-tectonic origin. Hot water drawn from depths of 1000–2000 m in the basin is a significant source of thermal energy in Hungary.

As mentioned earlier, Precambrian shields—be they in Australia, Canada, India, South Africa or the Ukraine—are characterized by low heat-flow values, about 30% less than average. It could be attributed to a relatively low concentration of radioactive elements in the underlying crust and upper mantle. According to the thermal convection hypothesis, stable shields are presumed to overlie a relatively cold upper mantle.

THERMAL HISTORY OF THE EARTH

The thermal history of the earth is closely linked to its structural evolution. Before the discovery of the radioactivity phenomenon, the outflow of heat from the earth was believed to be the result of the cooling of an initially hot body. Because the interior of the earth is hotter than its surface, it was vaguely imagined that at its origin the entire earth had been hotter and has subsequently cooled. If the hot-origin hypothesis were to be accepted, the earth, because of the subsequent radioactive heating, would have remelted instead of cooled.

The modern view is that the earth formed by accretion to be an initially cold body and that the primordial substance of the earth was similar in composition to that of a chondrite meteorite. Terrestrial material and chondrite show a similar isotopic composition in many respects. An argument against the chondrite model is the fact that the K/U ratio for chondrites is close to 6×10^4, while the value K/U $= 1 \times 10^4$ is more typical for terrestrial rocks Presumably the parent materials of the earth and the chondrites were similar and the more highly volatile potassium was lost during the formation of the earth to a much greater extent than in the formation of chondrites.

However, if we assume that the earth was formed by an accretion of chondritic substance at a low temperature about $4·5 \times 10^9$ years ago, we should be able to infer its subsequent thermal history. This can be solved mathematically by devising a number of models with assumptions regarding the initial temperature of the earth, its initial radioactive content (presumed to be distributed uniformly), its thermal conductivity, and so forth, and to calculate for each model what the present temperature distribution and heat flow would be. Such model calculations have been made (Lubimova, 1958; Macdonald, 1959) with the aid of electronic computers.

All the calculations indicate that, even if the earth's initial temperature had been low, the subsequent temperature increase by radioactive heating has been considerable. Fig.167 shows the results obtained by Lubimova; the computed temperature distribution is in reasonable agreement with the current esti-

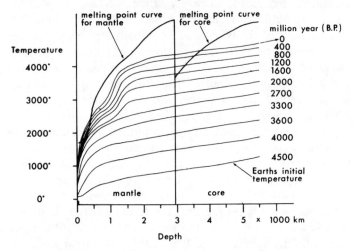

Fig.167. Thermal history of the earth in the geological past. Estimates are based on the assumption of a relatively cold primordial earth, subsequently heated by radioactivity. (Based on Lubimova, 1958.)

mates of temperature within the earth (see Fig.164). No allowance is made in this model for the lateral convective transport of heat.

Another interesting observation, first noticed by Birch (1965) and apparent in the figures of Table XIV, is the close equality between the measured terrestrial heat flow and the radioactive heat generation in an earth with an over-all composition of chondrites. What does this imply? If the hypothesis of the chondrite model is correct, the earth is losing as much heat as it is generating. The initially cold earth has been gradually increasing its temperature, but at present it seems to have approached a state of thermal balance. These implications sound very exciting, but in view of the uncertainties involved in various assumptions, they are to be treated as speculations. Birch (1965) gives an excellent discussion about the speculations on the earth's thermal history.

THERMAL PROSPECTING METHODS

In this section we consider the use of geothermic measurements for the investigation of near-surface and subsurface thermal sources. Local variations in temperature beneath the ground may be brought about by several causes: (1) the chemical actions resulting in exothermic reactions; (2) the presence of local radioactive heat sources; (3) the differences in conductivity of rocks; and (4) the presence of volcanic and hydrothermal sources.

For the investigation of thermal sources at great depth, the usual method involves heat-flow measurements over the area of interest. However, in the case of the exploration for shallow thermal sources, the measurements of temperatures at a depth of about 2 m are sufficient in most cases. This technique is relatively simple, measurements require little correction, and the interpretation of temperature anomalies is relatively straightforward. These features make it an attractive way of obtaining subsurface information for local use in addition to seismic and resistivity data.

In recent years thermal methods are being increasingly used in prospecting for groundwater, glacial and alluvial aquifers, thermal reservoirs, shallow salt domes, faults, and fissures, etc. An excellent discussion of the subject will be found in a monograph by Kappelmeyer (1974). It is possible here only to summarize the recent developments and to present a few examples.

Measuring techniques

Several precise thermistor units have been designed for field use. One of the designs (Krčkmář and Mǎsin, 1970) is able to measure from $0°$ to $70°C$, the relative accuracy in measurements being better than $0.01°C$. The sensing element is protected by a small case and may be connected to a metal rod with a length of 2–5 m or to a long cable if the measurements are to be made in deep boreholes.

The methods of measurements and their corrections have to take into account some disturbing effects of which the most important one is due to the diurnal variations of temperature caused by solar radiation. The variations may be observed in the soil down to a depth of a bout 1.5 m. In order to eliminate the effect of diurnal variations, the temperature measurements are made in shallow boreholes, drilled by a light drilling equipment or made by means of a steel rod and hammer. During the winter season, satisfactory results may be achieved by measuring beneath the snow cover, as indicated by Krčkmář (1968). To ensure a good contact between soil and sensor probe, the latter is pushed about 5 cm into the bottom of the hole; steady temperature is reached in most cases within half an hour.

Since it usually takes some weeks to cover a certain area, it is necessary to make corrections for the small regular shift in the annual temperature level. For this purpose, fixed sensors are kept at suitably located reference stations and readings are taken from them every day during a survey.

More recently infrared radiometers (IR) for airborne use have been developed in order to map the infrared imagery of large geothermal areas such as those existing in Iceland. Also meteorological satellites orbiting at high altitudes and equipped with high resolution devices (HRIR) have been used to record observations of infrared emission. Fig.168 shows the equivalent black-body radiation temperature profile from Greenland to Great Britain through Iceland as recorded by the Nimbus II satellite.

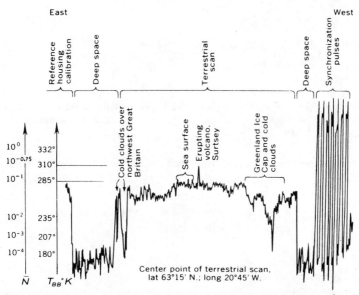

Fig.168. Effective radiance and equivalent black-body temperature (in degrees Kelvin) profile from Greenland to Great Britain through Iceland as recorded by the Nimbus II satellite. (After Friedman et al., 1969.)

EXAMPLES OF GEOTHERMAL SURVEYS

Sulphide ore deposits in Slovakia

This example is taken from a paper by Krčkmář and Măsin (1970). In the Spis-Gemer area of Slovakia, sideretic-sulphide

336

ores occur in crystalline schists of Palaeozoic age together with graphite beds. Generally the ores do not contain magnetic minerals, and their electrical conductivity is similar to that of the graphites. For this reason, both magnetic and electrical surveys do not yield satisfactory results. However, the exothermic reactions occurring in the oxidation zone of the sulphide deposits make it possible to apply geothermal investigations.

Fig.169 shows the temperature anomaly profiles observed over a part of the area. The rough topography and the changing

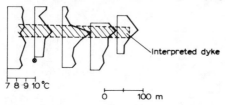

7 8 9 10 °C

0   100 m

Fig.169.Temperature anomalies (solid line) over sulphide deposits in Slovakia. The borehole is indicated by the crossed circle. (After Krčkmář and Măsin, 1970.)

vegetation cover caused differences in the temperature level and, therefore, only local anomalies higher than 1°C were marked. The zone of positive temperature anomalies in the eastern part of the area was examined in detail and checked by drilling, during which a vein with sulphidic mineralization was intersected.

Thermal water and hot vapour zones

In a number of cases thermal methods have been successfully applied to find fissures and cracks along which convective transfer of heat takes place from depth through the agency of water or gas. An example of the investigation of thermal water pockets in limestone cavities is given by Kappelmeyer (1961). Fig. 170 shows the geoisotherm map in an area of Neckertal (South Germany) where thermal water from a depth of 100–150 m seeps up through fissures and cracks and disturbs the local temperature distribution.

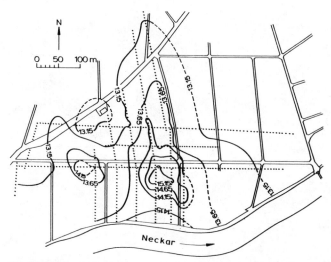

Fig.170.Geoisotherm map of an area in Neckertal (South Germany) based on temperature measurements at a depth of 1·5 m. Contour interval is 0·5°C. (After Kappelmeyer, 1961.)

When heat is transported by the seeping of gas or vapour, thermal anomalies observed at shallow depths could be more prominent. Temperature anomalies of a magnitude as large as 50°C have been recorded over some fissure zones on the island of Ischia in Italy (Kappelmeyer, 1961).

Salt and granite structures

Several studies have shown that a highly conductive salt intrusion causes a strong disturbance in the temperature of the overlying sediments.

An example of the thermal survey used to delineate salt structures is given by Poley and Van Steveninck (1970). Fig. 171 shows the temperature profile along a line about 10 km long, the maximum station spacing being 200 m. Included in the figure are the gravity data along the profile line, as well as the salt-layer thicknesses as encountered in the deep wells. The scale of the gravity profile has been inverted to show the

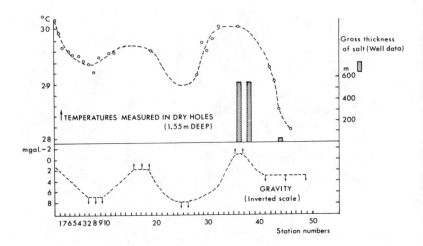

Fig.171.Similarity of gravity data (in mgal) and temperature profile over shallow salt structures. Spacing between stations is 200m. (After Poley and Van Steveninck, 1970.)

agreement more clearly. It is seen that a 1:1 correlation exists between a thermal high (maximum salt) and a gravity low. The salt distribution as found in the wells confirms the thermal predictions.

The example in Fig.172 is taken from Jakosky (1950). Here the temperature anomaly over a granite intrusion in contact with limestone is detectable at shallow depths. The anomaly curve resembles in shape that of a gravity profile expected over a fault structure. In the example shown, the temperature anomaly is attributable more to the higher radioactive heat generated in the granite rather than to its conductivity.

In this connection, mention may be made of some interesting observations of the temperature gradients obtained during the construction of the great tunnel, the St. Gotthard in Switzerland. During tunelling, when passing through gneisses and schists, the temperature gradient averaged 1°/47 m, whereas when traversing granite at the north end, the gradient

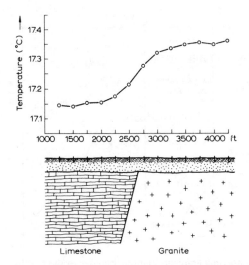

Fig.172.Shallow-temperature profile over a granite intrusion in contact with limestone. (After Jakosky, 1950.)

became 1°/21 m. From an analysis of rock specimens, it was found that the granite contained an unusually high proportion of radioactive material which would partly account for the higher temperatures recorded. In particular Rybach (1973) has emphasized that the heat production in the Swiss Alpine rocks is controlled entirely by their radioactive contents, and does not depend on variations in thermal conductivity.

Lithology information from temperature logs

The vertical gradient of the temperature in the earth varies within wide limits (5–70°C/km) depending upon the thermal properties of the rocks, and temperature logs in deep bore-holes can be used in correlating stratigraphic horizons.

As an example, Fig.173 shows the log of the temperature, the temperature gradient and the lithology in a borehole in West Texas. Significant variations in the temperature gradient are clearly indicated with changes in lithology. The greatest

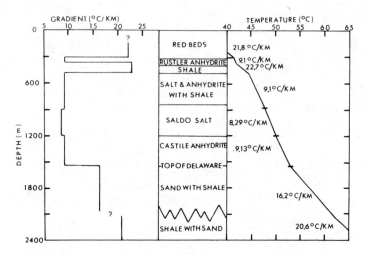

Fig.173.Temperature, geothermal gradients and the lithology of the North-rup well, Texas. (After Herrin and Clark, 1956.)

temperature gradient occurs in shales and the lowest in salt and anhydrite horizons. This is to be expected, since salt and anhydrite have a higher conductivity and consequently exhibit lower temperature gradients.

CHAPTER 8

Geophysics Applied to Global Tectonics

INTRODUCTION

The most striking feature of the earth, after its size and shape, is the division of its surface into continents and ocean basins. A satisfactory theory of tectonics should be able to explain not only the structural evolution of continents and oceans but also the source of energy required to maintain the observed motions of the different parts of the crust.

Have continents and oceans always been where they are today? Have they always had their present size and shape? What mechanism could account for their evolution? To these and many other similar questions, there are no definite answers as yet. This is mainly because the geological record of the relative motions of the outermost part of the earth is nowhere complete. Most of the available evidence of tectonic movements is circumstantial, and observations are generally open to several interpretations.

Until recently there were three main global theories to explain the earth's surface features:

(1) The earth is rigid and contracting: this theory assumes that the earth was hot at its origin and holds that the earth has been cooling since its beginning, that it became rigid at an early date, and that folding and thrusting of mountains is a result of internal compression caused by thermal contraction in the earth's interior. The discovery of radioactivity altered the original concept of the contracting earth without absolutely invalidating it. The contraction hypothesis still

persists, but it has lost favour among earth scientists, chiefly because it is unable to explain the widespread occurrence of tensional features such as block faulting and rift valleys.

(2) The earth is rigid and expanding: this theory speculates that the present distribution of continents could be explained by a large-scale expansion of the earth; the radius of the primitive earth being roughly one-half of its present value of about 6370 km. The theory is able to explain the occurrence of rifts in the ocean basins as tears in the crust resulting from expansion. As we shall see later, energy considerations and other available evidence rule out a large-scale expansion of the earth and, therefore, the expansion process alone is inadequate to account for the global tectonic features.

(3) The earth is mobile with the continents drifting over the mantle: the theory holds the view that the earth is slightly plastic with the continents slowly drifting over its surface, fracturing and reuniting and perhaps growing in the process. Since the 1920's the theory has been a subject of heated debate among earth scientists, dividing them into two groups which are currently referred to as the "mobilists" and "fixists". In contrast to the mobilists, the fixists hold the view that the earth has been rigid throughout its history, with fixed ocean basins and continents.

In the past two decades, knowledge about the physical features of the earth has vastly increased, particularly knowledge about the structure of the ocean floors, and many ideas have subsequently been transformed primarily as a result of important discoveries from bathymetric, seismic, magnetic, and heat-flow studies. These discoveries led to a general acceptance of the continental drift theory which, until the early 1950's, was severely criticized by many geologists and summarily rejected by several geophysicists. In the wake of its revival and general acceptance during the 1960's, the continental drift theory paved the way for the more sophisticated and revolutionary concepts of "sea-floor spreading" and "plate tectonics".

The purpose of this chapter is to present a brief account of the role which the recent geophysical discoveries played in transforming the older concepts of global tectonics into a unified science of plate tectonics which promises great advances in the field of earth sciences. The subject, although still in a state of development, has grown so vast that it requires collection and assimilation of an enormous amount of details which is beyond the scope of this review. For an extensive treatment of the subject, the reader is referred to the book by Le Pichon et al. (1973) and to a collection of papers edited by Cox (1973). A popular discussion of the subject can be found in Takeuchi et al. (1970), Dewey (1972), and Hallam (1973).

The successive discoveries and steps leading to the formulation and the subsequent development of the theory of plate tectonics contain many lessons for students. In the following sections only a few selected fragments of the total picture of global tectonics are considered in order to discuss their geophysical aspects.

We will begin by examining the evidence for recent crustal movements.

RECENT CRUSTAL MOVEMENTS

Relative motions in the upper part of the earth's surface involving vertical and horizontal deformation of the crust play a significant role in geotectonics. Vertical movements are easy to detect and until recently their importance was more widely recognized than that of horizontal displacements.

Precise geodetic and tide-gauge measurements give clear evidence that some areas of the crust (e.g. Fennoscandia) are at present slowly rising, whereas others (e.g. The Netherlands) are slowly sinking. The maximum rate of uplift (~ 10 mm/ year) known occurs in the central region of Fennoscandia (see Fig.63, p.123) and North Canada. Postglacial uplifts of Fennoscandia and Canada have been considered to be the result of isostatic rebound, and the attenuation of their uplifts with time has provided a means of estimating the viscosity of the upper mantle (see p.126).

There are, however, many areas whose uplift and subsidence cannot be accounted for solely by isostatic processes. Small (1963), for example, has indicated that in North America the velocity of relative movement in the eastern part is 3–5 mm/year; the Gulf Coast basin is subsiding while the Appalachian Mountains, the Ozark, and the Canadian shield are rising. In the western part, which is orogenic in character, the rate of movement is as great as 10–12 mm/year (the Rocky Mountains, the Great basin, and the Coast Ranges). On the Russian platform (Mescherikov, 1963) there are areas where recent movements do not correlate with the geological structure, on the contrary, they are of opposite character; that is, some basins are rising while arches are sinking. Analysis of these movements in the Russian platform suggests a block structure with maximum differential movements occurring at the boundaries between blocks.

Recent horizontal movements of the earth's crust have been studied in detail by many authors (e.g. Pavoni, 1964, 1966; Allen, 1965). Major horizontal movements between sections of the crust are noticeable on transcurrent faults of which the best known are the San Andreas fault, the Atacama fault (Chile), the Anatolia fault (Turkey), the Phillipine fault, and the Alpine fault (New Zealand). In several cases cumulative movements of hundreds of kilometres can be recognized with apparently the same direction of displacements over longer periods. The rate of displacement in big wrench-fault zones is estimated to be about 1–3 cm/year.

From tectonic analysis of horizontal displacements along active wrench-fault zones in the Alpide belt, Pavoni (1969) deduced the trajectories of the principal horizontal pressure (PHP) in the earth's crust (Fig.174). The agreement of these directions with those deduced from earthquake studies of stress patterns in the seismic active zones is rather striking (see Fig.175). This remarkable fact is an indication that the

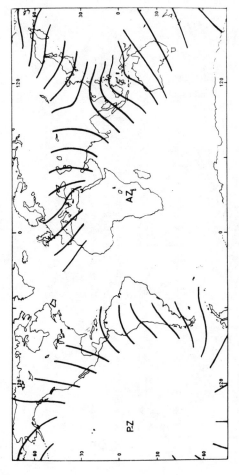

Fig.174.Directions of the principal horizontal pressure (PHP) in the Alpide belt as deduced from Cenozoic tectonics. Flow of mantle material away from the two geotectonic centres (PZ, AZ) below the lithosphere has been postulated to explain the present stress pattern. (After Pavoni, 1969.)

346

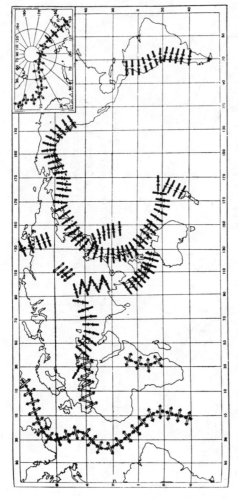

Fig.175.The orientation of principal stress axes in the seismic belts. The maximum relative compression is indicated by black arrows, the maximum relative tension by white arrows. (After Balakina et al., 1969.)

movements which led to the formation of young mountain belts continue today as evidenced by the seismicity and tectonicity of the belt areas.

In conclusion, we may say that there is ample evidence for recent crustal movements, both horizontal and vertical. Horizontal displacements are, as a rule, one order greater than vertical movements. The cause of these movements apparently lies in the earth's mantle. Before elaborating further on this point it is necessary to review the evidence for large-scale horizontal movements such as those associated with the drifting of continents.

CONTINENTAL DRIFT

The continental drift hypothesis has an exciting history. The hypothesis originated from the observation that the Atlantic coasts of Africa and South America show a remarkable similarity. The first comprehensive statement about continental drift was enunciated by the German meteorologist Alfred Wegener in 1912.

According to Wegener all the present continents were joined together in a single supercontinent ("Pangaea") during the upper Palaeozoic. During the Mesozoic and Tertiary, Pangaea broke into fragments and the continents moved westward or towards the equator or both. During the Cretaceous period, South America and Africa, like some gigantic iceberg that had split, began to drift apart, making room for the South Atlantic. Opening of the North Atlantic was mainly accomplished during the Pleistocene. Wegener's arguments, based on evidence collected from diverse disciplines such as geology, geophysics, palaeontology, and palaeoclimatology, are best described in the 1928 edition of his book *Die Entstehung der Kontinente und Ozeane.* One of the mainstays of his argument was the Permo-Carboniferous glaciation that affected South America, South Africa, Australia, and India, suggesting that these land masses were grouped around the south pole at that

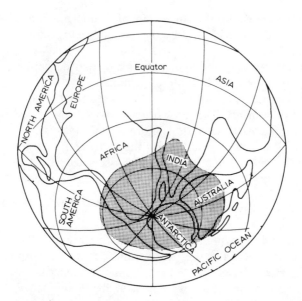

Fig.176.Distribution of the Permo-Carboniferous glaciation, with the continents reassembled according to Wegener's theory of continental drift. (Adopted from Holmes, 1944.)

time (Fig.176).

Since Wegener's original work, the continental drift hypothesis has been a subject of vigorous debate among earth scientists. Among the early supporters of Wegener's theory were the "mobilists" led by Holmes, Du Toit and Carey. The chief opponents were the "fixists", among whom the most formidable and steadfast has been the geophysicist Jeffreys, who, on energy considerations, strongly refuted Wegener's proposed mechanisms of tidal forces and the pole-fleeing centrifugal force ("Polflucht-Kraft").

A comprehensive account of the debate on continental drift, with successive discoveries and developments leading to the associated concepts of sea-floor spreading and plate tectonics, has been given by Wyllie (1971). The subject has also been discussed at a popular level in many books of which the one by Takeuchi et al. (1970) is probably the best, both in style and in presentation.

We shall present here only a few geophysically important aspects of the subject.

Fitting the continents together

The initial Wegenerian approach of matching the coastlines of the Atlantic has been placed on a quantitative basis by Bullard et al. (1965). They used a computer to determine the best fit of the Americas against Europe and Africa. The procedure was to select a depth contour (at 500 fathoms*) to define the continental margins, and to minimize the areas of misfit by rotational adjustments.

The South Atlantic fit turns out to be remarkably good, with only trivial overlaps and gaps (Fig.177). The fit of the North Atlantic continents, however, necessitates some significant adjustments. Spain has to be rotated anticlockwise with respect to the rest of Europe, otherwise it overlaps North Africa. It is also necessary to remove Iceland and other features of Tertiary age; on the other hand, to fill up the gap it is necessary to retain the Rockall Bank. All these adjustments seem to be justified by the presently available data. The reader will recall that there is independent palaeomagnetic evidence for the rotation of Spain (see p.256). Iceland and the dividing oceanic ridge are believed to be features younger than Palaeogene, and the Rockall Bank has recently been shown (Scrutton and Roberts, 1971) to have the seismic characteristics of continental crust. Thus, the fit of the northern continents, to form a single unit sometime before the Tertiary period, appears to be fairly convincing. In the same way the fit of ancient Gondwanaland pieces (i.e. the southern continents) around the Indian Ocean appears to be quite good (Smith and Hallam, 1970).

The next stage is to investigate the fit of geologic, tectonic,

* 1 fathom = 6 feet = 1·8 m.

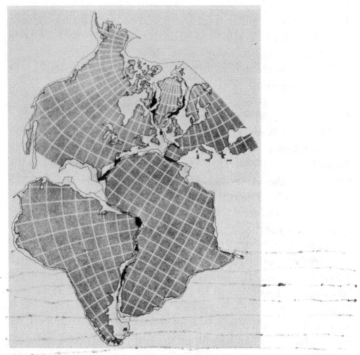

Fig.177. The "fit" of continents bordering the Atlantic Ocean at the 500-fathom depth line. Dark areas represent misfits. (After Bullard et al., 1965.)

and palaeontological features across the Atlantic and between the fragments of ancient Gondwanaland. These aspects have been examined in considerable detail by Hurley (1968) and Smith (1971), and their conclusions support the concept of two main continental masses in the upper Palaeozoic: Laurasia in the north and Gondwanaland in the south.

Palaeomagnetic evidence for the drift hypothesis

The subject of palaeomagnetism was introduced in Chapter 4 where are also briefly discussed the quantitative evidence provided by palaeomagnetic studies in support of the drift

hypothesis. It has been explained earlier (p.236) how palaeomagnetic measurements on rocks from a single continent enable
the determination of the pole positions for various geologic
periods, or, in other words, the construction of an apparent
polar wander curve for that particular continent.

Fig.178 shows the polar wandering curves for North America
and Europe. The pole relative to North America has apparently
followed a path from somewhere in the Pacific in late Palaeozoic times to its present position. There is, of course, no way
of determining from this information alone whether this
motion is due to the migration of the pole or to the motion
of the continent relative to the pole. The migration path of
the magnetic pole (coincident with the geographic pole according to the geocentric axial dipole hypothesis, see p.232)
is a global phenomenon, and should be independent of the site
of the observer. The fact that the polar wandering curves for

Fig.178.Polar wander curves for Europe (dotted line) and North America
(solid line.) (Redrawn from Runcorn, 1962).

North America and Europe do not agree is the essential palaeo-magnetic evidence for continental drift. Runcorn (1956) was the first to recognize that European and North American curves could be reconciled if the North Atlantic were closed by moving North America adjacent to Europe. This is consistent with the drift hypothesis that Europe and North America began to separate around the Jurassic.

Thus came the first valid palaeomagnetic evidence of past continental land movement, completely independent of evidence from other disciplines. Of course, this startling evidence from palaeomagnetism did not solve all the difficulties pertaining to the continental drift. Alternate interpretations have been proposed. Not every one was convinced about the statistical validity of the divergence between the two polar paths for Europe and North America. Some argued that the divergence in polar paths could result from secondary magnetization suffered by rock units on different continents. The assumption of a dipolar field prior to the Tertiary has also been challenged. These points have already been discussed at length in Chapter 4 where also pertinent explanations to refute the above objections are given.

Extension of palaeomagnetic studies to other continents revealed that each continent had a different polar wander path (see Fig.131). Whereas it is fairly easy to reconcile the two curves shown in Fig.178 by closing the Atlantic, the problem becomes much more complicated when so many curves are involved. Not only continental translation but also continental rotation has to be invoked. Often when data are scarce, the interpretation seems easy, but as the number of data grows, the interpretation becomes increasingly difficult. Palaeo-geographic reconstruction of land-masses from palaeomagnetic data is not a straightforward process and is often bound by several constraints; palaeolatitudes can be determined (see section *Palaeomagnetism and palaeolatitudes*, p.251) but not palaeolongitudes. For example, the relative positioning of Greenland between North America and Europe in the early Tertiary is still uncertain despite several palaeomagnetic studies (Tarling, 1967; Kristjansson and Deutsch, 1973; Athavale and Sharma, 1975).

Mechanisms for drift

The first postulated mechanism of continental drift was based on isostasy considerations (see p.118). This led to the concept of floating continents on the mantle, which, although certainly solid for most purposes, behaved as a fluid when subjected to stresses of long duration. This idea was not acceptable to most seismologists who felt convinced that, to a great depth, the earth was solid. Further, the forces which were invoked to move the continents, such as the tidal and the pole-fleeing centrifugal force, were uncomfortably small and mechanically unattractive (Jeffreys, 1970).

An alternative and more attractive mechanism considered was that the mantle undergoes large-scale slow convective motions, based on the assumption that the radioactive heat within the earth provides the energy source. Among the early proponents of the convection hypothesis, Holmes (1929) introduced the idea that convection currents in the solid mantle can cause both continental drift and mountain belts. Holmes envisaged that mantle convection would produce tension near the upwelling currents and compression near the sinking currents and between them (Fig.179). This early version of the convection hypothesis had at last outlined a plausible mechanism for drift. As we shall see later, this concept requires some modification to be in accord with modern ideas on ocean-floor spreading and plate tectonics. The evidence from several quarters, especially the evidence relating to the low-velocity channel in the upper mantle (see Fig.14, p.26), now very strongly supports the thought that, in spite of its solid character, the mantle is capable of undergoing convection, and that the continents could be transported as surface features on the convective cells. The mantle convection is not necessarily thermal, but may also be driven by density differences (see

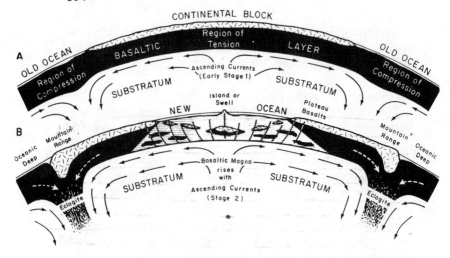

Fig.179.The convection hypothesis and Holmes' interpretation of continental drift. The upper diagram (A) shows the regions of tension and compression due to convective flow in the substratum. The lower diagram (B) shows the separation of a continental block with the formation of new ocean floors, mountain ranges, and ocean deeps. (After Holmes, 1944.)

Fig.191). Corroborative evidence for mantle motions is now appearing, particularly from geophysical studies of the ocean floors, and will be discussed at a later stage. Agreement has not been reached on the details of convection but the principle cannot be seriously disputed.

An alternative to the convection hypothesis that has been suggested is that the earth has undergone a radial expansion during geological time. The expanding earth hypothesis was originally suggested by Hilgenberg (1933). Carey (1958) adapted the expansion hypothesis because he found that the continents forming the Pangaea fitted much better on a globe of much smaller radius. He interpreted the post-Palaeozoic separation of the continents as the result of a rapid expansion of the earth, the Permian radius being about 0·75 times the present radius. Egyed (1957), on the other hand, advocated a slow expansion rate for the earth (about 0·05 cm/year in radius), basing his estimate on expansion presumably caused by phase transformations at the mantle–core boundary. More recently Dicke (1962) and Jordan (1967) have attributed expansion to the decrease of the gravitational constant, G, with time. Reviews of the expansion hypotheses are given by Irving (1964) and Holmes (1965). The palaeomagnetic results (Cox and Doell, 1961; Ward, 1963) are inconsistent with the hypothesis of Carey (1958) and Hilgenberg (1933), but not sufficiently precise to test the slow expansion rate proposed by Egyed. From energy requirements of an expanding earth, Beck (1961) precludes any large-scale expansion. It appears, therefore, that the expanding earth theory alone is inadequate to explain continental drift.

OCEAN-FLOOR TECTONICS

The floors of the oceans are simpler in structure and history than are the surfaces of the continents. By the 1950's seismic refraction measurements had firmly established the structure of the normal oceanic crust (see Table III, p.79), the main anomalous areas being the mid-ocean ridges and the deep trenches.

Of particular importance was the discovery, made from bathymetric studies, that the world-encircling mid-ocean-ridge system is a topographic rise, hundreds of kilometres in width and tens of thousands of kilometres in length. The ridge system is characterized by its seismicity (see Fig.17), a high heat flow (see Table XIII), local volcanicity, and axial rift, which imply tension. These diverse facts had somehow to be fitted together for an understanding of the evolution of the ocean floors.

Hess' concept of oceanic conveyor belt

It was Hess (1960, 1962) who first integrated the above facts about oceans into the hypothesis which we now know as "sea-floor-spreading". Expanding on Holmes' (1929) early version of convection currents, Hess postulated that the mid-ocean ridges are underlain by the hot, rising limbs of convection cells in the mantle, and that the sea floor is continuously created by a process of lateral accretion (spreading) away from the ridge crests. A significant feature of Hess' hypothesis is that the major part of the oceanic crust is essentially surfaced by a peridotite mantle partly hydrated to form serpentine at temperatures below 500°C. Hess maintained that oceanic layer 3 is nothing more than a surface expression of the mantle-convection process (Fig.180), with the mid-ocean ridges marking the sites of the rising limbs of the convection cells, and oceanic trenches being associated with the descending limbs of convection cells. Carried to its logical conclusion, the hypoth-

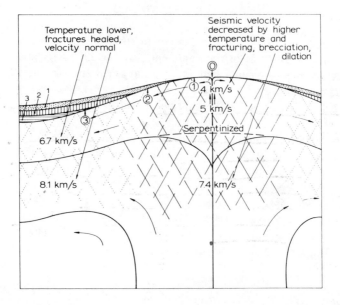

Fig.180. Hess interpretation of mid-oceanic ridge geology in terms of the serpentinization of the mantle and its lateral spreading away from the ridge crest. (After Hess, 1962, fig.7.)

esis suggests that this is the process by which continental drift occurs; the continents are carried passively on the tops of the mantle-convection cells rather than ploughing through the mantle, as proposed by the early "drifters"—and opposed by geophysicists (e.g. Jeffreys, 1970) on energy considerations.

Dietz (1961), as an alternative to Hess' hypothesis, suggested that oceanic layer 3* is mainly composed of gabbro and the upper mantle of eclogite (a high pressure form of gabbro?). Of particular importance is the observation that both Hess and Dietz considered the Moho as a boundary, not between two entirely different materials, but between two materials that are mutually transformable (peridotite–serpentine or eclogite–gabbro).

Magnetic record of sea-floor spreading

Spectacular support for the sea-floor spreading hypothesis has come from the study of oceanic magnetic anomalies. The sequence of discoveries and the steps that led Vine and Matthews (1963) to suggest an explanation of the strip-like pattern of oceanic magnetic anomalies, in terms of geomagnetic field reversals recorded in the ocean floor, have been already discussed in Chapter 4 and will not be repeated here. The reader will recall that according to the Vine-Matthews hypothesis, oceanic layer 2 (composed of basalt) has acted as a magnetic tape, recording the polarity of the magnetic field as new crust formed at the crest of the ridges. Thus Vine and Matthews suggested that Hess' conveyor belt of oceanic crust works also as a magnetic tape recorder.

The implications of this were tremendously exciting. If both sea-floor spreading and geomagnetic polarity reversals have occurred (see Fig.126), then we had a potential tape-deck

---

*The constitution of oceanic layer 3 is still under debate. Suggestions are that it is: (a) mainly serpentinite; (b) mainly metabasic rocks with amphibolites at the base; and (c) a mixture of various basic and ultrabasic plutonic and metamorphic rocks.

for determining the speed of the oceanic conveyor belt, given, of course, an accurate time scale for polarity reversals. The rate of spreading deduced from the magnetic polarity record appears to be of the order of a few centimetres per year (see Fig.128), although it varies from ridge to ridge, and along the ridge. Extrapolation backwards in time at these rates indicates an age for the ocean floor no older than Mesozoic, as Hess had indeed suggested.

The spreading rates for the past few million years are found to be more or less constant for a given area. In some specific areas, such as the South Atlantic, where no obvious hiatus in spreading is noticed, the spreading rate has been remarkably constant at 2 cm/year for either side of the ridge axis as far back as the late Cretaceous (80 m.y. B.P.). On this basis Heirtzler et al. (1968) were able to assign dates to geomagnetic field reversals extending far beyond the radiometrically established 4·5-m.y. scale of Fig.124. Fig.181 shows the geomagnetic reversal time scale of Heirtzler et al. as recently extended back to the base of the upper Jurassic (162 m.y. B.P.) by Larson and Pitman (1972). This bold extrapolation, back to at least 80 m.y., has stood a critical test of consistency as established by the JOIDES* sampling and microfossil dating of sediments directly overlying the basement across the South Atlantic at about 30°S (see Fig.130).

The drilling ship "Glomar Challenger" has in the past few years undertaken many cruises under an international programme known as DSDP (Deep Sea Drilling Project). The volume of data, accumulated from the study of several hundreds of drill cores, is already enormous and its analysis should keep many scientists busy for a long time. The oldest sediments yet discovered overlying oceanic layer 2 (basaltic basement) are middle to upper Jurassic, and occur in the northwest Pacific and the western North Atlantic. The core infor-

*Joint Oceonographic Institutes Deep Earth Sampling (a joint project of five American Universities).

mation from DSDP has indeed tended to confirm the sea-floor spreading theory; more than 50% of the present ocean floor appears to have been created in the past 70 m.y. (see Fig.129).

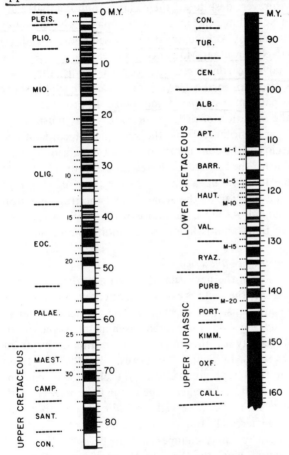

Fig.181.Geomagnetic polarity time scale as deduced from oceanic magnetic anomalies. From the present to 80 m.y. after Heirtzler et al. (1968); from 80 to 160 m.y. after Larson and Pitman (1972). Normal polarity zones are black, the reversed zones are white. Geological periods are indicated on the left, numbers assigned to prominent magnetic anomalies are given in the centre.

Seen in sharper detail the magnetic anomalies permit us to distinguish active oceanic ridges from inactive (extinct) ones by the presence of a pronounced central magnetic anomaly. However, despite all their clarity and compatibility with lateral spreading, the magnetic anomalies give us dates for the initiation of continental drift (by sea-floor spreading) in only relatively few areas. Many oceanic areas are magnetically quiet, anomaly-free zones (e.g. the western Pacific, the equatorial Atlantic, the northeastern Indian Ocean) which show no stripped pattern of magnetic anomalies. In order to assign an age to the initiation of drift in such areas one must resort to other criteria, notably from the geological record on the trailing margins of the continents and from the palaeomagnetic data available for the continents. Based on these considerations, it is possible to draw up a time-table for the drift (Fig.182) relating to the opening of the Atlantic and the fragmentation of Gondwanaland. As more and more of the sea-floor is dated, the suggested time-table is likely to be modified in details.

Subduction of the oceanic crust

The formation of the new crust at the crest of oceanic ridges implies that either the earth is expanding or equivalent amounts of old crust are consumed elsewhere. The former alternative is unlikely, judging from the present asymmetry in the distribution and activity of the ridges. The motion away from the ridge axes is dominantly east–west. Le Pichon (1968) computed, for different great circles, the changes in circumference which would take place if the ocean floor were created, in agreement with the observed spreading rates, but not consumed elsewhere. He found great differences between the circles, sufficient to imply changes in radii by as much as 500 km, which would cause the earth to be significantly distorted from its spheroidal shape. Thus, it seems preferable to assume that crustal material is somehow consumed at approximately the same rate at which new crust forms. Loci of sinking or downward motion may be regions of compression such as

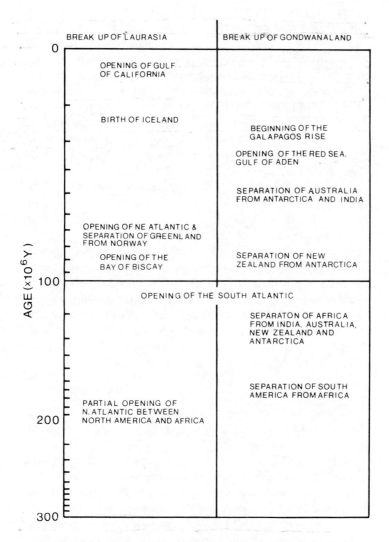

Fig.182.Suggested time scale for the opening of the Atlantic Ocean and fragmentation of Gondwanaland. (After Heirtzler, 1968, and Vine, 1971.)

young fold mountain belts, or the Benioff zones on which intermediate and deep earthquake foci lie. These zones, as will be recalled (see Fig.18), dip at approximately 45° below the continents or island arcs; their up-dip extension marks approximately the position of oceanic trenches.

Trenches, or fold mountain belts parallel to the coastline, are not found on both sides of the Atlantic, the only exception being the Puerto Rico trench. This suggests the absence of a downward motion of the sea floor on the margins of the Atlantic. Continents bordering it appear, therefore, to be carried or rafted along with the sea floor itself, thus accounting for continental drift and the progressive widening of the Atlantic Ocean. Alternatively, it is possible that one coast line of the Atlantic (for example, the eastern one) remains stationary while the axis of the ridge itself migrates westward at the same rate at which new crust spreads out on its eastern side.

The hypothesis of continuous creation of oceanic crust demands that accumulating sediments are continuously carried away by the moving crust. The thickness of sediments, and the age of the sediments should, therefore, gradually increase with distance from the axis of the nearest ridge. This is roughly what has been observed in the South Atlantic (see Fig.130), although the data elsewhere are not so consistent.

Against sea-floor spreading

Despite the large body of supporting evidence for the sea-floor spreading hypothesis, not all earth scientists have accepted the idea as indisputable.

Meyerhoff (1970) has marshalled pertinent geological and palaeoclimatological evidence that appears to be inconsistent with the idea of the closing and opening of the Atlantic during the past 800–1000 m.y. According to Meyerhoff and Teichert (1971), the distribution of glacial and coal deposits essentially requires certain moisture precipitation and ocean-current patterns, and these show that the Atlantic and Indian Oceans have existed since late Carboniferous or earlier time.

Beloussov (1970) has contested the validity of the evidence for sea-floor spreading on many grounds that range from subjectivity in the interpretation of marine magnetic anomalies to utter neglect of the known aspects of continental geology by adherents of the ocean-floor spreading hypothesis. He continues to believe in the idea that oceanic crust is formed by a "basification" of the continental crust, though the evidence cited for this remarkable process (Beloussov, 1968a, b) has only managed to convince few others.

As for Jeffreys, who has been a steadfast opponent of the continental drift hypothesis, he prefers not to mention sea-floor spreading at all in the newest edition of his book, *The Earth* (1970). The only comment he makes in passing is: "Magnetic anomalies on the Atlantic Ridge and off the coast of California have been interpreted as due to displacements of the order of 100 km. Whether this is correct or not, such displacements are of the order of those found in mountains, for which explanations exist; there is no reason to suppose it legitimate to extrapolate to displacements of whole continents for thousands of miles."

TRANSFORM FAULTS

A striking feature of ocean-ridges and trenches is their apparent offset in a number of places, apparently by faults along the transverse fracture zones. It will be recalled that magnetic surveys in the East Pacific revealed large offsets of magnetic anomaly pattern, such as shown in Fig.102. Along the Mendocino fracture zone the offset is much larger, amounting to more than 1000 km. Especially puzzling is the observation that these offsets associated with large-scale shear displacements seem to end abruptly along their length.

A simple yet profound explanation as to how these large strike-slip faults can terminate has been suggested, within the frame of the sea-floor spreading hypothesis, by Wilson (1965). He postulated that the fractures, which offset the ocean ridges and link island arc chains, are not simple transcurrent faults, but faults ("transforms") of a somewhat different character. In a transform fault the displacement is now active only in the segment lying between a pair of ridges where crust is formed, or a pair of island arc–trench systems, where crust is returning to the mantle. Fig.183 shows a ridge-ridge transform fault. Since new crustal material wells up at the ridge axis and moves outward on both sides of a fracture zone, it is clear that the material is moving in opposite directions along the zone BC between the offset sections of the axis.

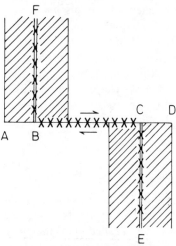

Fig.183.Schematic model of a ridge-ridge transform fault. Hatching indicates new surface area created during a given period of sea-floor spreading. Present seismic activity (indicated by crosses) is confined to ridge crests (BF and CE) and to segment BC of the fracture zone AD. Arrows denote sense of shear motion along active segment BC. (After Isacks et al., 1968.)

Supporting evidence for the transform fault hypothesis has come from seismology in two ways. First, the earthquake epicentres associated with the mid-ocean-ridge system are lined up on the crests of the ridges and on the parts of fracture zones that correspond to segment BC, not outside BC. Secondly, from the first motion studies of earthquakes (see P.39) along the fracture zones, the direction of the first motion is found to be just what one could predict on the basis of transform faulting. For instance, Fig.184 illustrates part of the Mid-Atlantic Ridge, showing the location of epicentres and the sense of slip given by first-motion studies of earthquakes on the transverse fracture zone segments. All earthquakes on the ridge-fracture systems are shallow-focus quakes, as could be predicted by sea-floor spreading processes.

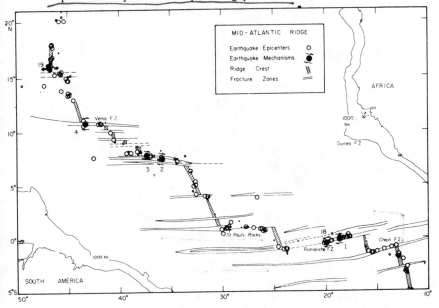

Fig.184. Location of epicentres (open circles) along the equatorial part of the Mid-Atlantic Ridge. Focal mechanism solutions for six earthquakes (solid dots numbered 1, 2, 3, 4, 18, 19) on the E–W fracture zones are shown. Sense of shear displacement and strike of inferred fault plane are indicated by arrows beside each of these mechanisms. (After Sykes, 1968, fig.10.)

The prediction and confirmation of transform faulting provided further support for the concept of sea-floor spreading.

PLATE TECTONICS (OR THE NEW GLOBAL TECTONICS)

The concept of plate tectonics has evolved from the hypotheses of sea-floor spreading and transform faults. The germinal idea that active belts consisting of ridges, trenches, and transform faults are connected in a continuous network which divides the earth's surface into several large, rigid "plates" was put forth by Wilson (1965). However, the theoretical formulation and development of the concept of plate tectonics is due to the work of four young scientists: McKenzie and Parker (1967), Morgan (1968), and Le Pichon (1968).

The concept of rigid plates

The basic idea is that the earth's rigid surface layer (lithosphere) suffers strong deformation only along relatively narrow linear mobile belts. These mobile belts, consisting of ridges, trenches, and the interconnecting transform faults, divide the lithosphere into a number of rigid, aseismic blocks ("plates") which do not suffer major internal deformation. Of particular importance is the relative motion between the plates, this motion is the fundamental cause of earth's tectonic activity.

Three types of plate boundaries are possible:

(1) Constructive or divergent junctures, where new crust is formed as plates move apart (crests of the mid-ocean ridges).

(2) Destructive or convergent junctures, where the crust is shortened (young fold mountain belts) as plates approach each other, or, consumed as one plate is thrust under the other (oceanic trenches).

(3) Conservative or shear junctures (or the transform faults), where plates slip laterally past each other, and the crust is conserved.

Where two plate boundaries, or three plates, meet is known as a triple junction, which is indeed the only way that the boundary between two rigid plates can end (see Fig. 187).

The geometry of the plate motion on a globe is deduced

as follows: if two blocks move apart, their relative motion can be described as a rotation of one block with respect to the other, or as a rotation of the two blocks in opposite sense. The pole of spreading or rotation (not to be confused with the entire earth's pole of rotation) corresponding to the motion that opens an oceanic ridge may be ascertained in two ways:

(1) Since the transform faults that intersect the ridge represent the direction of the spreading away from the ridge axis, all such faults must lie on small latitude circles centred on the pole of spreading centre A (Fig.185).

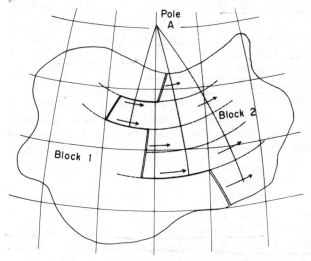

Fig.185.Geometrical relationship showing the rotation of rigid plates about a "pole of spreading". (After Morgan, 1968.)

(2) The rate of spreading at a point along the ridge must vary as the cosine of the angle of latitude (relative to A).

Wherever spreading rates, determined from the chronology of magnetic reversals, are sufficiently well known, the centres of rotation determined by these two methods agree rather well. The rates of the angular rotation of blocks found in this manner are of the order of $1°$ or less per million years.

Major plates of the earth's surface

To account for the global tectonic activity along the world-encircling system of seismic active belts, Le Pichon (1968) divided the earth's surface into six major plates, between which there are interactions as shown in Fig.186. In a twelve-plate model after Morgan (1971) six additional plates are named (shown hachured in Fig. 186). Note the close coincidence of the plate boundaries with the earthquake epicentre distribution (cf. Fig.17). The directions of relative motions at a few points are shown in Fig.186. The rates of the relative motions fall between 1 and 10 cm/year, thus giving spreading rates of 0·5–5 cm/year per ridge flank. By spreading rate we mean the half-rate of the separation of two plates, and assume that the spreading has been bilaterally symmetric.

Starting from these data on spreading rates it should be possible to go backward in geological time and infer the relative positions of the plates at some time in the past. This backward extrapolation (except for the recent past) can be hazardous because of the episodic character of sea-floor spreading, as indicated by the magnetic anomaly patterns observed over the ocean floors. The evolution of the Indian Ocean since the Cretaceous provides a pertinent example (McKenzie and Sclater, 1971). India apparently moved northwards from Antarctica at a high speed averaging over 14 cm/year between anomalies Nos. 31 and 22* (Fig.187). This rapid motion slowed down at about the time anomaly No.22 was formed and was followed by a period in which little or no sea-floor spreading took place west of the Ninety East Ridge (Fig.187). The short distance separating anomalies Nos.5 and 23 in Fig.187 are interpreted by McKenzie and Sclater as an almost complete

*Anomalies Nos.31 and 22 correspond to ages of about 72 m.y. and 56 m.y., respectively, on the Heirtzler et al. (1968) geomagnetic scale (see Fig.181).

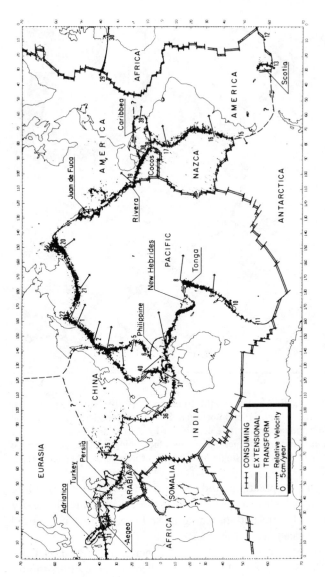

Fig.186.Present worldwide plate kinematic pattern at the surface of the earth. Earthquake epicentres (shown as small dots) roughly outline the present boundaries of the moving plates. In addition to the six large plates (Africa, America, Pacific, Antarctica, India, and Eurasia, used by Le Pichon, 1968), six small plates (shown hatched) are named after Morgan (1971). The vectors of differential motions are shown at selected points. (After Le Pichon et al., 1973.)

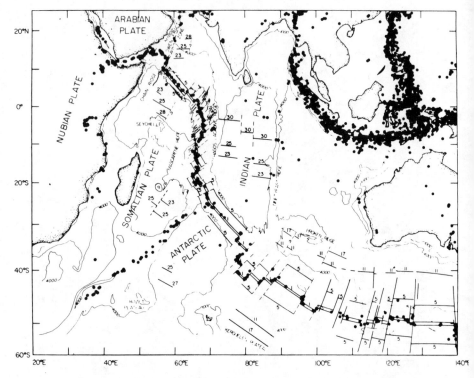

Fig.187. Spreading history of the Indian ocean. The Indian Ocean ridge system forms a triple junction between the Indian, Somalian, and Antarctic plates. Earthquake epicentres are shown by dots, ridge crests by double lines, and numbered magnetic anomalies (according to the geomagnetic time scale of Fig.181) by single lines. (After McKenzie and Sclater, 1971.)

cessation of spreading and subsequent change in spreading direction which was nearly north–south during the early episode. The present spreading episode seems to have started in the Miocene time (about 20 m.y. B.P.).

A comprehensive analysis of magnetic anomaly data from the Atlantic and Pacific Oceans has been made by Larson and Pitman (1972). Their estimates of average spreading rates deduced from motions across various plate boundaries are

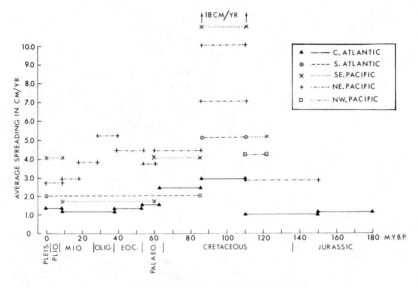

Fig.188.Average spreading rates in the Atlantic and Pacific Oceans as a function of geologic time. Note the marked increases in spreading rates of all spreading systems at 85–110 m.y. B.P. (After Larson and Pitman, 1972.)

shown in Fig.188. The spreading rate in the South Atlantic is found to be remarkably constant at 2 cm/year back to about 85 m.y. B.P. Note the marked increase in spreading rates of all spreading systems at 85–110 m.y. B.P.

Seismic evidence for the plate theory

Convincing support for the concepts of sea-floor spreading and plate tectonics is provided by seismology. Indeed, as is evident from Fig.17 and Fig.186, earthquakes occur almost exclusively on the presumed boundaries of plates where relative motions occur. Isacks et al. (1968) have determined the directions of the slip vectors from first-motion studies of shallow-focus earthquakes along the seismic active belts (Fig.189). These directions correspond remarkably well with

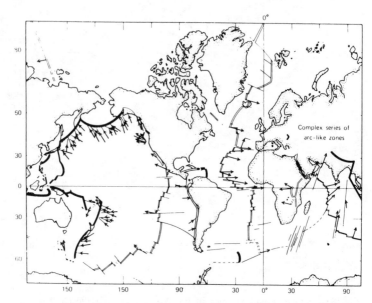

Fig.189.Summary map of slip vectors derived from earthquake-mechanism studies (see Fig.22 for the principle of the method). Arrows indicate the horizontal component of the direction of relative motion. (After Isacks et al., 1968.)

the movement directions of Fig.186 derived by Le Pichon from different evidence.

The best evidence for the thickness of the plates also comes from seismology. The velocity of the seismic waves is dependent on the density and elastic properties of the rock through which they pass. It will be recalled (see Fig.14) that the velocities of shear waves suddenly decrease below a surface about 70 km under the oceans and about 120 km under the continents. This suggests that an outer rigid layer about 70–120 km thick (the lithosphere), lying above a weaker and warmer layer (asthenosphere), probably constitutes the plates.

Models of lithosphere movements

Fig.190 is a block diagram illustrating schematically the

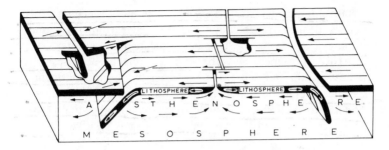

Fig.190. Block diagram illustrating schematically the configuration and role of the lithosphere (a layer of strength) in a version of the "new global tectonics". Arrows on top of the lithosphere indicate the relative movements of adjoining blocks. The arrows in the asthenosphere represent possible compensating flow in response to downward movement of lithosphere segments. (After Isacks et al., 1968.)

main features of plate tectonics or the "new global tectonics". The lithosphere is continuous except at the boundaries of the moving plates. The diagram shows the plates moving away from mid-ocean ridges and down into the asthenosphere at island arcs. Supporting evidence for the descent of lithosphere plates is provided by the occurrence of deep-focus earthquakes along dipping Benioff zones (see Fig.18). A return flow compensating for the downward movement of the lithosphere is shown in the asthenosphere with an uprise at the ridge crests. It will be appreciated that the terms "continental drift" and "sea-floor spreading" are now seen to be inappropriate in that individual plates typically include both continental and oceanic crust, although a few are entirely oceanic.

Since the continental crust is only about 35 km thick, whereas the plates are about 100 km or more thick, the continents ride as passengers on the plates. This fact answers one of the traditional objections to continental drift, namely the mechanical difficulty of having a continent ploughing its way across a strong rigid ocean floor. According to the plate tectonic view, continents and oceans are rafted along by the same conveyor belt. Nevertheless, continents, unlike oceans, impose certain important restraints on the plate motion. The narrow sharply defined trenches and the inclined earthquakes zones sloping away from the trenches indicate that oceanic crust is easily consumed by subduction, probably because it is relatively thin and dense. On the other hand, intracontinental seismic zones associated with mountain ranges exhibit compressional deformation over a wide area, which implies that continental crust is hard to consume because it is relatively thick, and buoyant.

A model of sea-floor spreading and subduction in petrological terms is illustrated diagrammatically in Fig. 191. The motion of partially molten mantle rock drags the lithospheric plate, presumably consisting of a thin layer of basalt on a thicker plate of peridotite, away from the ridge crest. The lithosphere slides on the seismic low-velocity layer which is plastic due to partial melting. Ultimately the oceanic crust, together with its coupled depleted ultramafic layer reaches an oceanic trench, where basalt is transformed to eclogite. The high density of dry eclogite ($\sim$3·5 g/cm^3) compared to that of mantle peridotite ($\sim$3·3 g/cm^3) causes gravitational instability and results in crustal subsidence. For further details about this speculative model, the reader is referred to Ringwood (1969).

Mountain building by plate movements

Plate tectonics are essentially ocean-based, i.e. the bulk of intense seismic and thermal activity is associated with the plate margins within the oceans and at the continental margins. More recent work, however, is directed towards the application of plate tectonics to continental geology.

As a crustal plate grows, its leading edge is destroyed at an equal rate. Sometimes the leading edge forcefully slides under the oncoming edge of another plate and returns to the asthenosphere. When this happens, a deep trench (such as the Tonga trench) in the Pacific is formed. In other areas where the leading edges of two plates come together (collide) at a relatively

375

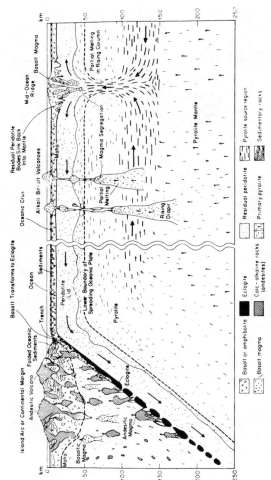

Fig.191.Diagrammatic sketch of a modified version of sea-floor spreading and subduction hypothesis. The depth of the return flow, shown by arrows in the middle of the figure, is unknown. (Based on Hess, 1962; reproduced from Ringwood, 1969, by permission.)

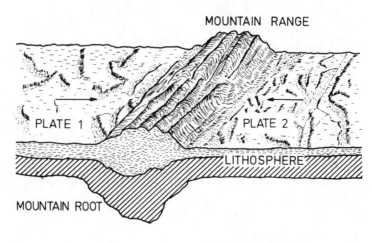

Fig.192.Mountain range is formed when the leading edges of two plates collide at a relatively slow speed (< 6 cm/year). The range consists of crustal material that folds under the compression exerted by the two plates. After Menard, 1969.

slow rate, the lithosphere crust can absorb the compression and buckle up into large mountain ranges (see Fig.192) such as the Himalayas. In these ranges folding and overthrusting deform and shorten the crust. A very informative article on the different types of mountain-building resulting from plate movements has been given by Dewey and Bird (1970).

Several crustal sinks are no longer active; however, their past can be deduced from their geology. Orogenic belts give significant clues to the creation and disappearance of oceans during geological time. The Ural and Appalachian–Caledonian mountain belts, which lie within ancient Pangaea, have narrow ophiolite* complexes. These old ophiolite zones, like those in the Alpine–Himalayan belt, mark the sites of vanished oceans. Thus, Hamilton (1970) has presented a convincing case for the Ural Mountains being a late Palaeozoic line of closure between Europe and Siberia. Others (e.g. J.T. Wilson, 1966; Dietz, 1972) have attempted to explain the formation of the Appalachian fold belt on the hypothesis that the Atlantic Ocean has opened, closed, and reopened.

The foregoing citation of some examples should by no means leave an impression that in plate tectonics we necessarily have found the final key to the understanding of mountain building. There are a number of problems, particularly of older orogenies, which pose real difficulties for the plate theory. No one has yet satisfactorily accounted for the late Palaeozoic Hercynian orogeny in western Europe. Many mountain belts older than 600 m.y. do not have ophiolitic complexes like those of the younger mountain belts; the presence of these belts is difficult to explain by plate tectonics as it is understood today. Moreover, not all mountain belts, as marked by zones of compression and uplift, necessarily form at the plate margins.

Mid-plate tectonics

The theory of plate tectonics has explained the location and cause of most seismicity, volcanism and active mountain building, all of which are associated with the plate margins. There are, however, several striking exceptions where such tectonic activities occur away from the plate margins. A notable example is the chain of Hawaiian islands where extensive volcanism has created a mountain range near the centre of one of the largest plates. Such "mid-plate" phenomena can also occur in continental regions, as evidenced by continental grabens and active seismic zones. For example, in the

*Ophiolite complexes consist of submarine pillow basaltic lavas associated with serpentinized peridotite and radiolarian cherts. The lavas may show oceanic tholeiite geochemistry, or they may be sodic.

United States a highly active seismic zone is centred near New Madrid. These and other phenomena may account for a relatively small proportion of the earth's tectonic activity, but they are by no means insignificant and can hardly be dismissed as local.

Wilson (1963) attributed the island volcanicity of Hawaii to the rise of magmas from a nearly stationary "hot spot" in the upper mantle. The westward motion of the Pacific plate over the magma source resulted in the formation of the island chain which becomes progressively older from east to west. Morgan (1971) and Wilson (1973) have further developed the hot-spot theory. According to this theory, hot spots, marked by volcanism, a high heat flow, and uplift, are the surface expressions of deep-seated plumes rising in the mantle (Fig.193). In particular Morgan (1971) has suggested that the convective plumes spread out radially in the asthenosphere to drive the lithospheric plates.

Driving mechanisms for plate motions

The force which drives the plates is still unknown and it is likely to remain a speculative matter for a considerable time. This is because of our ignorance of the mantle, which is inaccessible to direct observation. Sea-floor spreading involves the formation of new crust along a ridge, while old crust moves down somewhere into the mantle. Clearly, some sort of convection is required to link the lithosphere and asthenosphere in a simple source-to-sink system.

The simple model of thermal convection within the mantle (Fig.179), with the migration of continents or ocean floor being coupled directly to the upper limbs of the underlying convective cells, is inadequate to account for the mechanics of the plates. For instance, it is difficult to envisage how both the rising and descending limbs of the convection cells can be abruptly offset by transform faults. The geometry of the ridge-ridge transform faults in the equatorial Atlantic would imply convection cells that are implausibly narrow

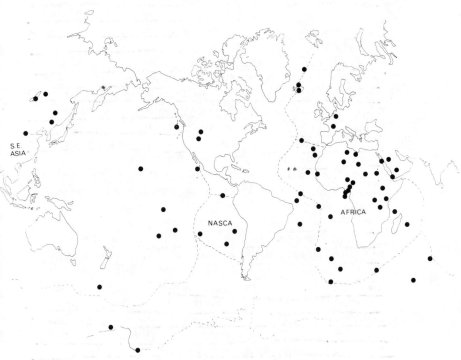

Fig.193.Sketch map of the world showing the location of supposed plumes rising from deep in the mantle. Hot spots on the earth's surface, marked by volcanism, high heat flow, and uplift, are the surface expressions of deep-seated plumes in the mantle. (After Wilson, 1973.)

compared to their length and depth extent.

Also it is difficult to see how the simple cell model can explain the behaviour of the African plate (Fig.186) which for the most part is bounded by the Mid-Atlantic Ridge and the ridge in the Indian Ocean. This plate is growing on both sides, and since there is no intermediate trench the two ridges must be moving apart. It is hard to imagine that the rising limbs of the convecting cells could keep exact pace with the moving ridges.

Alternative models have been suggested for the driving force or forces. The cold descending slab of the lithosphere (Fig. 191) is a major source of gravitational energy, and it may pull the rest of the plate after it (Elsasser, 1971). Density contrasts with the surrounding asthenosphere will cause sinking and the sea floor will be pulled along in its wake, with mantle material welling up to fill the gaps at the ridge axis. That the "sinks" act as stress guides once they have been initiated seems very likely, but the hypothesis fails to account for their initiation, and it seems doubtful if motion could be maintained for long. The situation is further complicated by the continuous migration of the "sinks" and "sources" at the surface which implies that the pattern of convective flow changes continuously.

Another speculative yet attractive model is that involving "hot spots" and "plumes" in the mantle which we mentioned previously (p.378). In particular, Morgan (1971) claims that the mantle "hot spots" can provide the driving force for plate motions. One of his arguments is that the gravity pattern and regionally high topography around each hot spot suggest that more than just surface volcanism is involved at each hot spot. A recent world gravity map (Kaula, 1970) made from satellite-orbit studies, shows regional gravity highs over Iceland, Hawaii, and most of the other hot spots. No physical theory has yet been given, but Morgan believes that one may be possible.

Apparently there is no shortage of possible driving mechanisms: plates may be pushed apart by mantle upwelling at the ridges, pulled apart by sinks beneath the trench systems, carried on a convection conveyor belt involving ascending hot and descending cold plumes, or slide under the influence of gravity. It is also possible that some small plates are mechanically driven by the effects of relative motion between adjacent large plates. The difficulty lies in devising critical tests to decide between the different hypotheses of driving mechanisms and causative forces.

In relation to plate movements, however, the basic question as to whether the lithosphere is an active or a passive element in the dynamics is yet unsettled. Does material rise along a ridge because that is where plates move apart and provide an opening, or is it the rising material that forces the plates apart? Some other related questions are: What is the role of phase changes in promoting or restraining convective motions? In what way do the stresses, necessary for earthquakes, originate in the down-sinking lithospheric plate? What is the physical theory of the supposed mantle plumes providing the driving force for plate motions? These are major questions which remain challenging problems in global tectonics.

APPENDIX

The International System of Units (SI)

The International System of Units, denoted SI (Système Internationale), was established by the General Conference of Weights and Measures in 1960 as a comprehensive electrical-mechanical-thermodynamical system of units. Since then it has been adopted by many scientific and technical agencies, including the National Aeronautics and Space Administration, the Council of the Royal Society (London), the European Association of Exploration Geophysicists and the International Association of Geomagnetism and Aeronomy.

The SI is an extension of the MKSA (metre-kilogramme second-ampere) system. The purpose of forming the SI was to end the confusion due to the existence of numerous systems of units, particularly in the fields of electricity and magnetism, and to provide a system which includes units which are actually used in measurements, such as ampere, volt, ohm, and watt. No c.g.s. system includes these units.

Additional information on the use of the SI, including tables of physical and geophysical constants expressed in SI-units, will be found in an article by Markowitz (1973).

Table XV lists some SI-units which are frequently used in geophysics. The corresponding equivalents in c.g.s. (or electromagnetic c.g.s.) units are also given.

TABLE XV
Conversion of SI-units to c.g.s. or electromagnetic c.g.s units

quantity	SI units name	symbol	Corresponding equivalent in c.g.s. or electromagnetic c.g.s. units
Mass	kilogramme	kg	10^3 g
Length	metre	m	10^2 cm
Time	second	s	s
Acceleration	metre/second^2	m/s^2	10^2 gal $= 10^2$ cm/s^2
Subunit for gravity	gravity unit	g.u. $= \mu$m/s^2	10^{-1} milligal (mgal)
Density	kilogramme/ metre^3	kg/m^3	10^{-3} g/cm^3
Force	newton	N	10^5 dynes
Pressure	pascal	Pa $=$ N/m^2	10 dynes/cm$^2 =$ 10^{-5} bar
Energy	joule	J	10^7 ergs $\simeq 0 \cdot 24$ cal
Power	watt	W $=$ J/s	10^7 ergs/s
Heat flow	watt/metre^2	W/m^2	$23 \cdot 9$ μ cal/cm^2 s
Conductivity (thermal)	watt/metre $^\circ$C	W/m $^\circ$C	$2 \cdot 39 \times 10^{-3}$ cal/ cm s $^\circ$C
Current	ampere	A	10^{-1} e.m.u. (or "absolute amp")
Potential difference	volt	V	10^8 e.m.u.
Electric field	volt/metre	V/m	10^6 e.m.u.
Electric charge	coulomb	C $=$ A s	10^{-1} e.m.u.
Capacitance	farad	F $=$ C/V	10^{-9} e.m.u.
Resistance	ohm	$\Omega =$ V/A	10^9 e.m.u.
Resistivity	ohm metre	Ωm	10^{11} e.m.u.
Magnetic flux	weber	Wb $=$ V s	10^8 maxwell
Magnetic flux density (B-field)	tesla	T $=$ Wb/m^2	10^4 gauss (G)
Subunit for the B-field	nanotesla	nT	$1\gamma = 10^{-5}$ gauss (G)
Magnetic field strength (H-field)	ampere/metre	A/m	$4\pi\ 10^{-3}$ oersted (Oe)
Inductance	henry	H $=$ Wb/A	10^9 e.m.u.
Permeability*	henry/metre	$\mu_0 = 4\pi\ 10^{-7}$ H/m	1 (for vacuum)
Susceptibility	dimensionless	k	4π e.m.u.
Magnetic pole strength	ampere metre	A m	10 e.m.u.
Magnetic moment	ampere metre^2	A m^2	10^3 e.m.u.
Magnetization	ampere/metre	A/m	10^{-3} e.m.u.

*Permeability for vacuum.

References

Abrahamsen, N., 1967. Some archaeomagnetic investigations in Denmark. *Prospez. Archaeol.*, 2:95–97.

Ade-Hall, J.M., 1964. The magnetic properties of some submarine oceanic lavas. *Geophys. J.R. Astron, Soc.*, 9:85–92.

Affleck, J. (Editor), 1964. Special issue on magnetic methods. *Geophysics*, 29(4):481–623.

Affleck, J. (Editor), 1965. Special issue on magnetic methods. *Geophysics*, 30(5):705–891.

Ahorner, L., 1970. Seismotectonic relations between the Graben zones of the upper and lower Rhine Valley. *Int. Upper Mantle Project. Sci. Rep.*, No. 27, pp.155–166.

Ahorner, L., Murawski, H. and Schneider, G., 1972. Seismotektonische Traverse von der Nordsee bis zum Apennin. *Geol. Rundsch.*, 61 : 915–942.

Airy, G.B., 1855. On the computation of the effect of the attraction of mountain masses, as disturbing the apparent astronomical latitude of stations in geodetic surveys. *Phil. Trans. R. Soc., Lond., Ser. B*, 145:101–104.

Aitken, M.J., 1961. *Physics and Archaeology*. Interscience, New York, N.Y., 181 pp.

Allan, T.D., 1969. Review of marine geomagnetism. *Earth Sci. Rev.*, 5:217–254.

Alldredge, L.R., Van Hooris, G.D. and Davis, T.M., 1963. A magnetic profile around the world. *J. Geophys. Res.*, 68:3679–3692.

Allen, C.R., 1965. Transcurrent faults in continental areas. *Phil. Trans. R. Soc. Lond., Ser. A*, 258:82–89.

Allingham, J.W. and Zietz, I., 1962. Geophysical data on the Climax Stock, Nevada test site. *Geophysics*, 27:599.

Alpin, L.M., Berdichevksi, M.N., Vedrintsev, G.A and Zagarmistr, A.M., 1966. *Dipole Methods for Measuring Earth's Conductivity*. Consultants Bureau, New York, N.Y., 302 pp.

Åm, K., 1972. The arbitrarily magnetized dyke: interpretation by characteristics. *Geoexploration*, 10:63–90.

Anderson, D.L., 1967. Latest information from seismic observations. In: T.F. Gaskell (Editor), *The Earth's Mantle*. Academic Press, London, pp.355–420.

386

Anstey, N.A., 1970. *Seismic Prospecting Instruments, 1 (Geoexploration Monograph 3).* Gebruder Borntraeger, Berlin, 156 pp.

Athavale, R.N. and Sharma, P.V., 1975. Palaeomagnetic results on Early Tertiary lava flows from West Greenland and their bearing on the evolution history of the Baffin Bay–Labrador Sea region. *Can. J. Earth Sci.,* 12: 1–18.

Athavale, R.N., Hansraj, A. and Verma, R.K., 1972. Palaeomagnetism and age of Bhander and Rewa sandstones from India. *Geophys. J. R. Astron. Soc.,* 28:499–509.

Baker, P.E., 1957. Density logging with gamma rays. *Am. Inst. Min. Metall. Engrs. Tech. Publ.,* No.4654. Also: *J. Pet. Technol.,* 9(10).

Balakina, L.M., Misharina, L.A., Shirokova, E.I. and Vendenskaya A.V., 1969. The field of elastic stresses associated with earthquakes. In: P. J. Hart (Editor), *The Earth's Crust and Mantle (Geophysical Monograph 13).* American Geophysical Union, Washington, D.C., pp. 166–171.

Balsley, J.R. and Buddington, A.F., 1960. Magnetic susceptibility anisotropy and fabric of some Adirondack granites and orthogneisses. *Am. J. Sci., Bradley Volume,* 258A:6–20.

Barazangi, M. and Dorman, J., 1969. World seismicity map of ESSA, Coast and Geodetic Survey epicenter data for 1961–1967. *Bull. Seismol. Soc. Am.,* 59:369–380.

Bates, R.G., 1966. Airborne radioactivity surveys, an aid to geological mapping. In: *Mining Geophysics, 1.* Society of Exploration Geophysicists, Tulsa, Okla., pp.67–76.

Båth, M., 1954. Seismicity of Fennoscandia and related problems. *Gerlands Beitr. Geophys.,* 63:173–206.

Båth, M., 1961. Die Conrad–Diskontinuität. *Freiberger Forschungsh.,* C 161, 34 pp.

Båth, M., 1966. Earthquake energy and magnitude. *Phys. Chem. Earth,* 7:115–165.

Båth, M., 1968. *Mathematical Aspects of Seismology.* Elsevier, Amsterdam, 415 pp.

Båth, M., 1971. Average crustal structure of Sweden. *Pure Appl. Geophys.,* 88:75–91.

Båth, M., 1973. *Introduction to Seismology.* Birkhäuser Verlag, Basel, 395 pp.

Bean, R.J., 1953. Relation of gravity anomalies to the geology of central Vermont and New Hampshire. *Bull. Geol. Soc. Am.,* 64:509–538.

Beck, A.E., 1961. Energy requirements of an expanding earth. *J. Geophys. Res.,* 66:1485–1490.

Beloussov, V.V., 1968a. Some general aspects of development of the tectonosphere. *23rd Int. Geol. Congr., Prague, 1968*, 1 : 9–17.

Beloussov, V.V., 1968b. An open letter to J. Tuzo Wilson. *Geotimes*, 13(10): 17–19.

Beloussov, V.V., 1970. Against the hypothesis of ocean-floor spreading. *Tectonphysics*, 9:489–511.

Benioff, H., 1964. Earthquake source mechanisms. *Science* 143: 1399–1406.

Bentz, A., 1961. *Lehrbuch der Angewandten Geologie, 1*. Ferdinand Enke Verlag, Stuttgart, 1071 pp.

Bhattacharya, B.K. and Morley, L.W., 1965. The delineation of deep crustal magnetic bodies from total field aeromagnetic anomalies. *J. Geomag. Geoelectr.*, 17:237–252.

Bhattacharya, P.K. and. Patra, H.P., 1968. *Direct Current Geoelectric Sounding*. Elsevier, Amsterdam, 135 pp.

Bhimasankaram, V.L.S., 1964. Partial self-reversal in pyrrhotite. *Nature*, 202:478–479.

Bhimasankaram, V.L.S. and Rao, B.S.R.,1958. Mangenese ore of southern India and its magnetic properties. *Geophys. Prospect.*, 6:11–24.

Birch, F., 1952. Elasticity and constitution of the earth's interior. *J. Geophys. Res.*, 57:227–289.

Birch, F., 1954. Heat from radioactivity. In: H. Faul (Editor), *Nuclear Geology*. John Wiley and Sons, New York, N.Y., pp.148–174.

Birch, F., 1961. Composition of the earth's mantle. *Geophys. J.R. Astron. Soc.*, 4: 295–311.

Birch, F., 1965. Speculations on the earth's thermal history. *Bull. Geol. Soc. Am.*, 76:133–154.

Birkenmajer, K., Krs, M. and Nairn, A.E.M., 1968. A palaeomagnetic study of Upper Carboniferous rocks from the Inner Sudetic Basin and the Bohemian Massif. *Bull. Geol. Soc. Am.*, 79:589–608.

Blackett, P.M.S., 1961. Comparison of ancient climates with the ancient latitudes deduced from rock magnetic measurements. *Proc. R. Soc. Lond., Ser. A.*, 263:1–30.

Blackett, P.M.S., Clegg, J.A. and Stubbs, P.H.S., 1960. An analysis of rock magnetic data. *Proc. R. Soc. Lond., Ser. A.*, 256:291–322.

Blohm, E.K. and Flathe, H., 1970. Geoelectrical deep sounding in the Rhinegraben. *Int. Upper Mantle Project. Sci. Rep.*, No. 27, pp.239–242.

Blundell, D.J. and Read, H.H., 1958. Palaeomagnetism of the younger gabbros of Aberdeenshire and its bearing on their deformation. *Proc. Geol. Assoc. (England)*, 69:191–203.

Boigk, H. and Schöneich, H., 1970. Dei Tiefenlage der Permbasis in nordlichen Teil des Oberrheingraben. *Int. Upper Mantle Project. Sci. Rep.*, No.27, pp.45–55.

Books, K.G., White, W.S. and Beck, M.E., Jr., 1966. Magnetization of Keweenawan gabbro in northern Wisconsin and its relation to time of

388

intrusion. *U.S. Geol. Survey Prof. Paper*, No.550-D:117-124.

Bott, M.H.P., 1956. A geophysical study of the granite problem. *Q.J.Geol Soc. Lond.*, 112:45–67.

Bott, M.H.P., 1961. The granite layer. *Geophys. J. R. Astron. Soc.*, 5: 207–216.

Bott, M.H.P., 1971. *The Interior of the Earth*. Edward Arnold, London, 316pp.

Bott, M.H.P. and Smithson, S.B., 1967. Gravity investigations of subsurface shape and mass distributions of granite batholits. *Bull. Geol. Soc. Am.*, 78:859–878.

Bouguer, P., 1749. *La Figure de la Terre*. Paris, 365 pp.

Bowen, R. 1966. *Palaeotemperature Analysis*. Elsevier, Amsterdam, 265 pp.

Briden, J.C., 1967. Recurrent continental drift of Gondwanaland. *Nature*, 215:1334–1339.

Briden, J.C., 1970a. Evidence of complexity of the ancient geomagnetic field and some possible implications. *J. Earth Sci. (Leeds)*, 8:115–123.

Briden, J.C., 1970b. Palaeolatitude distribution of precipitated sediments. In: S.K. Runcorn (Editor), *Palaeogeophysics*. Academic Press, London, pp.437–444.

Brownell, G.M., 1950. Radiation surveys with a scintillation counter. *Econ. Geol.*, 45:167–174.

Brune, J.N., 1969. Surface waves and crystal structure. In: P.J. Hart (Editor), *The Earth's Crust and Upper Mantle (Geophysical Monograph 13)*. American Geophysical Union, Washington, D.C., pp.230–242.

Brynjolfsson, A., 1957. Studies of remanent magnetism and viscous magnetism in the basalts of Iceland. *Phil. Mag.*, 6:247–254.

Bucha, V., 1965. Results of archaeomagnetic research in Czechoslovakia, *J. Geomag. Geoelectr.*, 17:407–412.

Bullard, E.C., 1971. The earth's magnetic field and its origin. In: I.G. Gass, P.J. Smith and R.C.L. Wilson (Editors), *Understanding the Earth*. Artemis Press, Sussex, pp.71–78.

Bullard, E.C., Freedman, C., Gellman, H. and Nixon, J., 1950. The westward drift of the earth's magnetic field. *Phil. Trans. R. Soc. Lond. Ser. A.*, 243:67–92.

Bullard, E.C., Maxwell, A.E. and Revelle, R., 1956. Heat flow through the deep sea floor. *Adv. Geophys.*, 3:153–181.

Bullard, E.C., Everett, J.E. and Smith, A.G., 1965. The fit of the continents around the Atlantic. In: P.M.S. Blackett, E. Bullard and S.K. Runcorn (Editors), *Symposium on Continental Drift—Phil. Trans. R. Soc. Lond., Ser. A*, 258:41–51.

Bullen, K.E., 1963. *An Introduction to the Theory of Seismology*. Cambridge University Press, Cambridge, 3rd ed., 381 pp.

Cann, J.R., 1968. Geological processes at mid-ocean ridge crests. *Geophys. J. R. Astron. Soc.*, 15:331–341.

Carey, S.W., 1958. The tectonic approach to continental drift. In: S.W. Carey (Editor), *Continental Drift, a Symposium*. University of Tasmania, Hobart, pp.177–355.

Carlsborg, H., 1963. Om gruvkompasser, malmletning och kompassgångare. *Med Hammare och Fackla (Stockholm)*, 23:9–108.

Chandra, U., 1970. Comparison of focal mechanism solutions obtained from P- and S-wave data. *J. Geophys. Res.*, 75:3411–3430.

Chevallier, R., 1925. L'aimantation des lavas de l'Etna et l'orientation du champ terrestre en Sicile du 12e et 17e siècle. *Ann. Phys.*, 4:5–162.

Clark, Jr., S.P. (Editor), 1966. *Handbook of Physical Constants. Geol. Soc. Am., Mem.*, No. 97, rev. ed., 587 pp.

Clark, Jr. S.P. and Ringwood, A.E., 1964. Density distribution and constitution of the mantle. *Rev. Geophys.*, 2:35–88.

Closs, H., 1969. Explosion seismic studies in Europe. In: P.J. Hart (Editor), *The Earth's Crust and Upper Mantle (Geophysical Monograph 13)*. American Geophysical Union, Washington, D.C., pp.178–188.

Collette, B.J., 1968. On the subsidence of the North Sea area. In: D.T. Donovan (Editor), *Geology of the Shelf Seas*. Oliver and Boyd, London, pp.15–26.

Collinson, D.W., Creer, K.M. and Runcorn, S.K. (Editors), 1967. *Methods in Palaeomagnetism*. Elsevier, Amsterdam, 609 pp.

Compagnie Generale de Geophysique, 1963. *Master Curves for Electrical Sounding*. European Association of Exploration Geophysicists, The Hague, 2nd ed., 49 pp.

Conrad, V., 1925. Laufzeitkurven des Taurnbebens vom 28. November 1923. *Mitt. Erdbeben Komm., Wien*, No.59, 23 pp.

Cook, A.H., 1973. *Physics of the Earth and Planets*. Macmillan, London, 316 pp.

Cook, A.H. and Murphy, T. 1952. Measurements of gravity in Ireland. Gravity survey of Ireland north of the line Sligo-Dundalk. *Dublin Inst. Adv. Studies, Geophys. Mem.*, No.2, Part 4, pp.1–36.

Corbato, C., 1965. Thickness and basal configuration of lower Blue Glacier, Washington, determined by gravimetry. *J. Glaciol.*, 5:637–650.

Cox, A., 1969. Geomagnetic reversals. *Science*, 163:237–245.

Cox, A. (Editor), 1973. *Plate Tectonics and Geomagnetic Reversals*. W.H. Freeman and Company, San Francisco, Calif., 702 pp.

Cox, A. and Doell, R.R., 1961. Paleomagnetic evidence relevant to a change in the Earth's radius. *Nature*, 189:45–47.

Cox, A. and Doell, R.R., 1962. Magnetic properties of the basalt in hole EM 7, Mohole Project. *J. Geophys. Res.*, 67:3997–4004.

Crain, I.K., 1971. Possible direct causal relation between geomagnetic reversals and biological extinctions. *Bull. Geol. Soc. Am.*, 82:2603–2606.

Creer, K.M., 1970. A review of palaeomagnetism. *Earth Sci. Rev.*, 6: 369–466.

Creer, K.M. and Sanver, M., 1967. The use of the sun compass. In: D.W. Collison, K.M. Creer and S.K. Runcorn (Editors), *Methods in Palaeomagnetism*. Elsevier, Amsterdam, pp.11–15.

Creer, K.M., Irving, E. and Nairn, A.E.M., 1959. Paleomagnetism of the Great Whin sill. *Geophys. J. R. Astron., Soc.*, 2:306–323.

Currie, R.G., Gromme, C.S. and Verhoogen, J., 1963. Remanent magnetization of some upper Cretaceous granitic plutons in the Sierra Nevada, California. *J. Geophys. Res.*, 68:2263–2279.

Dansgaard, W., Johnsen, S.J., Möller, J. and Langway, C.C., 1969. One thousand centuries of climatic record from Camp Century on the Greenland ice sheet. *Science*, 166:377–381.

Dearnley, R., 1966. Orogenic fold-belts and a hypothesis of Earth evolution. *Phys. Chem. Earth*, 7:1–114.

Delaney J.P., 1940. Leonardo da Vinci on isostacy. *Science*, 91:546.

Deutsch, E.R., 1965. Palaeolatitudes of Tertiary oil fields. *J. Geophys. Res.*, 70:5193–5203.

Deutsch, E.R., 1966. The rock magnetic evidence for continental drift. In: G.D. Garland (Editor), *Continental Drift—R. Soc. Can., Spec. Publ.*, No.9, pp.28–52.

Deutsch, E.R., 1969. Palaeomagnetism and North Atlantic paleogeography. *Am. Assoc. Petrol. Geol., Mem.*, No. 12, pp.931–954.

Dewey, J.F., 1972. Plate tectonics. *Sci. Am.*, 226:56–68.

Dewey, J.F. and Bird, J.M., 1970. Mountain belts and the new global tectonics. *J. Geophys. Res.*, 75:2625–2647.

Dicke, R.H., 1962. The earth and cosmology. *Science*, 138:653–664.

Dietz, R.S., 1961. Continent and ocean basin evolution by spreading of the sea floor. *Nature*, 190:854–857.

Dietz, R.S., 1972. Geosynclines, mountains, and continent building. *Sci. Am.*, 226:30–38.

Dix, C.H., 1952. *Seismic Prospecting for Oil*. Harper, New York, N.Y., 414 pp.

Dix, C.H., 1966. Seismic prospecting. In: S.K. Runcord (Editor), *Methods and Techniques in Geophysics*. Interscience, London, pp.249–278.

Dobrin, M.B., 1960. *Introduction to Geophysical Prospecting*. McGraw-Hill, New York, N.Y., 435 pp.

Doell, R.R. and Cox, A., 1961. Palaeomagnetism. *Adv. Geophys.*, 8:221–313.

Dorman, J., 1969. Seismic surface-wave data on the upper mantle. In: P.J. Hart (Editor), *The Earth's Crust and Upper Mantle (Geophysical*

391

Monograph 13). American Geophysical Union, Washington, D.C., pp.257-265.

Egyed, L., 1957. A new dynamic conception of the internal constitution of the earth. *Geol. Rundsch.*, 46:101-221.

Elder, J.W., 1968. Convection—the key to dynamic geology. *Sci. Progr.*, 56:1-33.

Elsasser, W.M., 1958. The earth as a dynamo. *Sci. Am.*, 198:44-48 (Reprint No. 825).

Elsasser, W.M. 1966. . Thermal structure of the upper mantle and convection. In: P.M. Hurley (Editor), *Advances in Earth Sciences.* M.I.T. Press, Cambridge, Mass., pp.461-502.

Elsasser, W.M., 1971. Sea-floor spreading as thermal convection. *J. Geophys. Res.*, 76:1101-1112.

Emiliani, C., 1955. Pleistocene temperatures. *J. Geol.*, 63:538-578.

Ewing, J., 1969. Seismic model of the Atlantic Ocean. In:P.J. Hart (Editor), *The Earth's Crust and Upper Mantle (Geophysical Monograph 13).* American Geophysical Union, Washington, D.C., pp.220-225.

Fabiano, E.B. and Peddie, N.W., 1969. Grid values of total magnetic field intensity, IGRF-1965. *U.S. Coast Geodet. Survey, E.S.S.A. Tech. Rep.*, No. 38, 55 pp.

Fahrig, W.F., Gaucher, E.H. and Larochelle, A., 1965. Palaeomagnetism of diabase of the Canadian shield. *Can. J. Earth. Sci.*, 2:278-298.

Faust, L.Y., 1951. Seismic velocity as a function of depth and geological time. *Geophysics*, 16:192-206.

Flathe, H., 1955. A practical method of calculating geoelectrical model graphs for horizontally stratified media. *Geophys. Prospect.*, 3:268-294.

Friedman, J.D., Williams, Jr., R.S., Palmason, G. and Miller, C.D., 1969. Infrared surveys in Iceland—preliminary report. *U.S. Geol. Survey Prof. Paper*, No. 650-C:89-105.

Fuller, M.D., 1962. A magnetic fabric in till. *Geol. Mag.*, 99:233-237.

Garland, G.D., 1965. *The Earth's Shape and Gravity.* Pergamon Press, London, 183 pp.

Garland, G.D., 1971. *Introduction to Geophysics—Mantle, Core and Crust.* W.B. Saunders, Philadelphia, Ill., 420 pp.

Gass, I.G. and Masson-Smith, D., 1963. The geology and gravity anomalies of the Troodos massif, Cyprus. *Phil. Trans. R. Soc. Lond., Ser. A*, 255:417-467.

Gassmann, F., 1961. Solution of an *n*-layer problem by seismic reflection method. *Geophys. J. R. Astron. Soc.*, 4:151-157.

Gassmann F., 1972. *Seismische Prospektion.* Birkhäuser Verlag, Basel,

392

430 pp.

Gassmann F. and Weber, M., 1960. *Einführung in die Angewandte Geophysik*. Hallwag, Bern, 284 pp.

Geodetic Reference System, 1967. Publ. Spec. Bur. Centrale Assoc. Int. Geod., Paris 1971, 116 pp.

Gilbert, W., 1600. *De Magnete*. London.

Gilvarry, J.J., 1957. Temperatures in the earth's interior. *J. Atmos. Terr. Phys.*, 10:84–95.

Girdler, R.W., 1964. Geophysical studies of rift valleys. *Phys. Chem. Earth*, 5:121–156.

Gough, D.I., 1956. A study of the paleomagnetism of the Pilansberg Dykes. *Mon. Not. R. Astron. Soc. Suppl.*, 7:196–213.

Graham, J.W., 1949. The stability and significance of magnetism in sedimentary rocks. *J. Geophys. Res.*, 54:131–167.

Graham, J.W., 1954. Magnetic susceptibility anisotropy, an unexploited petrofabric element. *Geol. Soc. Am. Bull.*, 65:1257–1258.

Graham, K.W.T., 1961. The remagnetization of a surface outcrop by lightning currents. *Geophys. J. R. Astron. Soc.*, 6:85–102.

Granar, L., 1958. Magnetic measurements on Swedish varved sediments. *Arkiv Geofys.*, 3:1–40.

Grant, F.S. and West, G.F., 1965. *Interpretation Theory in Applied Geophysics*. McGraw-Hill, New York, N.Y., 583 pp.

Gregersen, S., 1970. Surface-wave dispersion and crust structure in Greenland. *Geophys J. R. Astron. Soc.*, 22:29–39.

Griffiths, D.H. and King, R.F., 1965. *Applied Geophysics for Engineers and Geologists*. Pergamon, London, 223 pp.

Gurwitsch, I.I., 1970. *Seismische Erkundung*. Akademische Verlagsgesellschaft, Leipzig, 699 pp.

Gutenberg, B., 1959. *Physics of the Earth's Interior*. Academic Press, New York, N.Y., 240 pp.

Haalck, H., 1953. *Lehrbuch der Angewandten Geophysik, 1*. Gebrüder Borntraeger, Berlin, pp.27–155.

Habberjam, G., 1966. A nomogram for the investigation of the three-layer refraction problem. *Geoexploration*, 4:219–225.

Haddon, R.A.W. and Bullen, K.E., 1969. An earth model incorporating free earth oscillation data. *Phys. Earth Planet. Interiors*, 2:35–49.

Hahn, A., 1971. Types of magnetic anomalies measured on land and general aspects of their geological meaning. *I.A.G.A. Bull.*, 28:134–143.

Hahn, A. and Zitzmann, A., 1969. The relation of magnetic anomalies to topography and geologic features in Europe. In: P.J. Hart (Editor), *The Earth's Crust and Upper Mantle (Geophysical Monograph 13)*. American Geophysical Union, Washington, D.C., pp.399–404.

Hallam, A., 1973. *A Revolution in the Earth Sciences: From Continental Drift to Plate Tectonics*. Oxford University Press, Oxford, 127 pp.

Hamilton, N. and Rees, A.I., 1970. The use of magnetic fabric in palaeocurrent estimation. In: S.K. Runcorn (Editor), *Palaeogeophysics*. Academic Press, London, pp.445–464.

Hamilton, W., 1970. The Uralides and the motion of the Russian and Siberian platforms. *Bull. Geol. Soc. Am.*, 81:2553–2576.

Hammer, S., 1939. Terrain corrections for gravimeter stations. *Geophysics*, 4:184–209.

Hammer, S., 1963. Deep gravity interpretation by stripping. *Geophysics*, 28:369–378.

Hanus, V. and Miroslav, K., 1963. Palaeomagnetic dating of hydrothermal deposits in Czechoslovakia. *Geophys. J. R. Astron. Soc.*, 8:82–101.

Harland, W.B., Smith, A.G. and Wilcock, B. (Editors), 1964. *Geological Society Phanerozoic Time Scale—Q. J. Geol. Soc. Lond.*, 120-S:260–262.

Harris, P., 1971. The composition of the earth. In: *Understanding the Earth*. Artemis Press, Sussex, pp. 53–70.

Hart, P.J. (Editor), 1969. *The Earth's Crust and Upper Mantle (Geophysical Monograph 13)*. American Geophysical Union, Washington, D.C., 735 pp.

Heirtzler, J.R., 1968. Sea-floor spreading. *Sci. Am.*, 219:60–70.

Heirtzler, J.R., Le Pichon, X. and Baron, J.G., 1966. Magnetic anomalies over the Reykjanes Ridge. *Deep-Sea Res.*, 13:427–443.

Heirtzler, J. R., Dickson, G. O., Herron, E. M., Pitman, W. C. and Le Pichon, X., 1968. Marine magnetic anomalies, geomagnetic field reversals, and motions of the ocean floor and continents. *J. Geophys. Res.*, 73: 2119–2136.

Heiskanen, W. A. and Vening Meinesz, F. A., 1958. *The Earth and its Gravity Field*. McGraw-Hill, New York, N.Y., 470 pp.

Hermance, J. F., 1973. An electrical model for the sub-Icelandic crust. *Geophysics*, 38: 3–13.

Hermance, J. F. and Garland, G. D., 1968a. Magnetotelluric deep-sounding experiments in Iceland. *Earth Planet. Sci. Lett.*, 4:469–474.

Hermance, J. F. and Garland, G. D., 1968b. Deep electrical structure under Iceland. *J. Geophys. Res.*, 73:3797–3800.

Herrin, E. and Clark Jr., S. P., 1956. Heat flow in West Texas and Eastern New Mexico. *Geophysics*, 21:1087–1089.

Hess, H. H., 1960. Evolution of ocean basins. *Rep. Off. Naval Res., ONR Contract*, No. 1858(10).

Hess, H. H., 1962. History of ocean basins. In: A.E.J. Engel, H.L. James and B. F. Leonard (Editors), *Petrologic Studies: A Volume to honour A. F. Buddington*. Geological Society of America, pp. 599–620.

394

Hilgenberg, O.C., 1933. *Vom wachsendem Erdball*. Geissmann and Bartsch, Berlin.

Hill, M. N., 1957. Recent geophysical exploration of the ocean floor. *Phys. Chem. Earth*, 2:129–163.

Holmes, A., 1929. Radioactivity and earth movements. *Trans. Geol. Soc. Glasgow*, 18:559–606.

Holmes, A., 1944. *Principles of Physical Geology*. Thomas Nelson, London, 532 pp.

Holmes, A., 1947. The construction of a geologic time scale. *Trans. Geol. Soc. Glasgow*, 21:117–152.

Holmes, A., 1965. *Principles of Physical Geology*. Nelson, London, 2nd ed., 1288 pp.

Holtzscherer, J. J., 1954. Mesures seismiques. In: *Contribution a la Connaissance de l'Inlandsis du Groenland, Expeditions Polaires Francaises*. Missions Paul-Emile Victor, Paris, pp. 7–26.

Homilius, J., 1969. Geoelectrical investigations in East Afghanistan. *Geophys. Prospect.*, 17: 468–487.

Honkasalo, T., 1960. On the uplift of land in Fennoscandia. *Geophysica*, 7(2): 117–119.

Hospers, J. and Van Andel, S. I., 1969. Palaeomagnetism and tectonics, a review. *Earth. Sci. Rev.*, 5: 5–44.

Hospers, J. and Van Andel, S. I., 1970. A review of selected palaeomagnetic data from Europe and North America and their bearing on the origin of the North Atlantic Ocean. In: S. K. Runcórn (Editor), *Palaeogeophysics*. Academic Press, London, pp. 263–276.

Howell, L. G., Heintz, K. O. and Barrs, A., 1966. The development and use of a high pressure downhole gravity meter. *Geophysics*, 31:764–772.

Hubbert, M. K., 1932. Results of earth resistivity survey on various geologic structures in Illinois. *Am. Inst. Min. Metall. Engrs., Tech. Publ.*, No. 463, 23 pp.

Hummel, J. N., 1932. A theoretical study of apparent resistivity in surface potential methods. *Trans. Am. Inst. Min. Metall. Engrs.*, 97: 392–422.

Hurley, P. M., 1968. The confirmation of continental drift. *Sci. Am.*, 218: 52–64.

Innes, M. J. S., 1960. Gravity and Isostasy in northern Canada and Manitoba. *Publ. Dominion Observ., Ottawa*, 21 (6): 263–335.

International Atomic Energy Agency, 1969. *Nuclear Techniques and Mineral Resources.*, I.A.E.A., Vienna, 546 pp.

Irving, E., 1964. *Paleomagnetism and its Applications to Geological and Geophysical Problems*. John Wiley and Sons, New York, N.Y., 399 pp.

Irving, E. and Green, R., 1957. The paleomagnetism of the Kainozoic basalts of Victoria, *Mon. Not. R. Astron. Soc. Geophys. Suppl.*, 7: 347–359.

Isacks, B., Oliver, J. and Sykes, L. R., 1968. Seismology and the new global tectonics. *J. Geophys. Res.*, 73: 5855–5899.

Iszac, I. G., 1965. A new determination of nonzoned harmonics by satellites. In: G. Veis (Editor), *Proceedings of the Second International Symposium on the Use of Artificial Satellites for Geodesy, Athens, 1965.* 2: 223–229.

Ito, H., 1964. Paleomagnetic study on Kyushu outer zone. 1963 *Annu. Progr. Rep. Rock Magnetism Res. Group in Japan*, 112 pp.

Jacobs, J. A., Russel, R. D. and Wilson, J.T., 1959. *Physics and Geology.* McGraw-Hill, New York, N.Y., 424 pp.

Jahren, C. E., 1963. Magnetic susceptibility of bedded iron-formation. *Geophysics*, 28: 756–766.

Jahren, C. E., 1965. Magnetization of Keweenawan rocks near Duluth, Minnesota. *Geophysics*, 30: 858–874.

Jakosky, J. J., 1950. *Exploration Geophysics.* Trija, Los Angeles, Calif., 1015 pp.

Jeffreys, H., 1939. The times of P, S and SKS and the velocities of P and S. *Mon. Not. R. Astron. Soc. Geophys, Suppl.*, 4: 498–533.

Jeffreys, H., 1970. *The Earth.* Cambridge University Press, Cambridge, 535 pp.

Jeffreys, H. and Bullen, K. E., 1967. *Seismological Tables.* British Association of Advances in Science, London, 50 pp.

Jensen, H., 1970. *Lectures on Solid Earth Physics.* Inst. Geophys., University of Copenhagen, Publ. No. 2, 200 pp. (in Danish).

Jordan, P.,1967. *Die Expansion der Erde.* Fried, Vieweg and Sohn, Braunschweig, 180 pp.

Jung, K., 1961. *Schwerkraftverfahren in der Angewandten Geophysik.* Geest and Portig, Leipzig, 348 pp.

Kaila, K. L. and Narain, H., 1970. Interpretation of seismic refraction data and the solution of the hidden layer problem. *Geophysics*, 35: 613–623.

Kane, M. F., 1962. A comprehensive system of terrain corrections using a digital computer. *Geophysics.* 27: 455–462.

Kappelmeyer, O., 1961. Geothermik, In: A. Bentz (Herausgeber), *Lehrbuch der Angewandten Geologie.* Ferdinand Enke Verlag, Stuttgart, pp. 863–888.

Kappelmeyer, O., 1974. *Geothermics with Special Reference to Methods of Application.* Gebrüder Borntraeger, Berlin, 234 pp.

396

Kaula, W. M., 1970. Earth's gravity field: relation to global tectonics. *Science*, 169: 982 984.

Kawai, N,, Ito, H. and Kume, S., 1961. Deformation of the Japanese islands as inferred from rock magnetism. *Geophys. J. R. Astron, Soc.*, 6: 124 130.

Keller, G. V., 1971. Natural-field and controlled source methods in electromagnetic exploration. *Geoexploration*, 9: 99 147.

Keller, G. V., and Frischknecht, F. C., 1966. *Electrical Methods in Geophysical Prospecting*. Pergamon Press, London, 517 pp.

Kent, P. E., 1968. Geological problems in North Sea exploration. In: D. T. Donovan (Editor), *Geology of the Shelf Seas*. Oliver and Boyd, London, pp. 73-91.

Khan, M. A., 1960. The remanent magnetization of the basic Tertiary igneous rocks of Skye. *Geophys. J. R. Astron. Soc.*, 3: 45 62.

Khan, M. A., 1962. The anisotropy of magnetic susceptibility of some igneous and metamorphic rocks. *J. Geophys. Res.*, 67: 2873 2885.

Khattri, K., 1973. Earthquake focal mechanism studies—a review. *Earth Sci. Rev.*, 9: 19 63.

Khramov, A. N. and Sholpo, L. E., 1967. *Palaeomagnetism*. Nedra Press. Leningrad, 233 pp.

King, R. F., 1955. The remanent magnetism of artificially deposited sediments. *Mon. Not. R. Astron. Soc. Geophys.*, *Suppl.* 7: 115 134.

Klitten, K., 1972. Geoelectric mapping of postglacial basins. *Inst. Appl. Geol. Copenhagen, Mem.*, No. 1, 74 pp. (in Danish.)

Knopoff, L., 1967. Thermal convection in the earth's mantle. In: T. F. Gaskell (Editor), *The Earth's Mantle*. Academic Press, London, pp. 171 196.

Kobayashi, K., 1959. Chemical remanent magnetization of ferromagnetic minerals and its application to rock magnetism. *J. Geomag. Geoelectr.* 10: 99 117.

Koefoed, O., 1968. *The Application of the Kernel Function in Interpreting Geoelectrical Resistivity Measurements*. Gebrüder Borntraeger, Berlin, 111 pp.

Kosminskaya, I. P., Belyaevsky, N. A., and Volvovsky, I. S., 1969. Explosion seismology in the U.S.S.R. In: P. J. Hart (Editor), *The Earth's Crust and Upper Mantle (Geophysical Monograph 13)*. American Geophysical Union, Washington, D.C., pp. 195 208.

Kovach, R., 1966. Seismic surface waves: some observations and recent developments. *Phys. Chem. Earth*, 6: 251 314.

Kovach, R. L. and Breiner, S., 1967. A search for the piezomagnetic effect along the San Andreas Fault. *U.S. Progr. Rep. Upper Mantle Project, Natl. Acad. Sci. Natl. Res. Council, U.S.A.*, p.77.

Krčkmář, B., 1968. Anwendung der Geothermik bei der gelogischen Prospektion. *Freiberger Forschungsh.*, C 338: 45–33.

Krčkmář, B. and Mäsin J., 1970. Prospecting by the geothermic method. *Geophys. Prospect.*, 18: 255–260.

Kristjansson, L. G. and Deutsch, E. R., 1973. Magnetic properties of rock samples from the Baffin Bay Coast. In: P. J. Hood (Editor), *Earth Science Symposium on Offshore Eastern Canada—Geol. Survey Can. Paper*, 71-23: 573–598.

Kuhnen, F. and Furtwängler, P., 1906, Bestimmung der absoluten Grosse der Schwerkraft zu Potsdam. *Veröff. Preuss, Geodat. Inst.*, No. 27, 397 pp.

Kulp, J. L., 1961. Geological time scale. *Science*, 133: 1105.

Kunetz, G., 1966. *Principles of Direct Current Resistivity Prospecting*. Gebrüder Borntraeger, Berlin, 103 pp.

LaCoste, L. J. B., 1967. Measurement of gravity at sea and in the air. *Rev. Geophys.*, 5: 477–526.

Lang, A. H., 1970. Radioactivity methods. In: *Prospecting in Canada— Geol. Survey Can.*, Econ. Geol. Rep., No. 7, pp.151–159.

Larochelle, A., 1966. Palaeomagnetism of the Abitibi dyke swarm. *Can. J. Earth Sci.*, 3: 671–683.

Larson, R. L. and Pitman III, W. C., 1972. World-wide correlation of Mesozoic magnetic anomalies and its implications. *Bull. Geol. Soc. Am.*, 83: 3645–3662.

Lee, W. H. K. (Editor), 1965. *Terrestrial Heat Flow (Geophysical Monograph 8)*. American Geophysical Union, Washington, D.C., 276 pp.

Lehmann, I., 1967. Low-velocity layers. In: T. F. Gaskell (Editor), *The Earth's Mantle*. Academic Press, London, pp. 41–60.

Le Pichon, X., 1968. Sea-floor spreading and continental drift. *J. Geophys. Res.*, 73: 3661–3697.

Le Pichon, X., Francheteau, J. and Bonnin, J., 1973. *Plate Tectonics*. Elsevier, Amsterdam, 300 pp.

Libby, W. F., 1955. *Radiocarbon Dating*. University of Chicago Press, Chicago, Ill., 124 pp.

Liebscher, H. J., 1964. Deutungsversuche für die Struktur der tieferen Erdkruste nach reflexionsseismischen und gravimetrischen Messungen im Deutschen Alpenvorland. *Z. Geophys.*, 30: 51–96.

Lögn, O., 1954. Mapping nearly vertical discontinuities by earth resistivities. *Geophysics*, 19: 739–760.

Loncarevic, D. B., Masón, C. S. and Matthews, H. D., 1966. Mid-Atlantic Ridge near 45 north, I. The Median Valley, *Can. J. Earth Sci.*, 3: 327–349.

Lovborg, L., 1972. Assessment of uranium by gamma-ray spectrometry. In: S. H. U. Bowie, M. Davis and D. Ostle (Editors), *Uranium Pros-*

398

pecting Handbook. Institution of Mining and Metallurgy, London, pp. 157–173.

Lubimova, E. A., 1958. Thermal history of the earth with consideration of the variable thermal conductivity of the mantle. *Geophys. J. R. Astron. Soc.*, 1: 115–134.

Lubimova, E. A., 1967. Theory of thermal state of the earth's mantle. In: T. F. Gaskell (Editor), *The Earth's Mantle*. Academic Press, London, pp. 232–323.

Macdonald, G. J. F., 1959. Calculations on the thermal history of the earth. *J. Geophys. Res.*, 64: 1967–2000.

Macdonald, G. J. F., 1965. Geophysical deductions from observations of heat flow. In: W. H. K. Lee (Editor), *Terrestrial Heat Flow (Geophysical Monograph 8)* American Geophysical Union, Washington, D.C., pp. 191–210.

Markowitz, W., 1973. SI, the international system of units. *Geophys. Survey*, 1:217–241.

Mason, R.G., 1958. A magnetic survey of the west coast of the United States between 32°N and 36°N, 121°W and 128°W. *Geophys. J. R. Astron. Soc.*, 1:320–329.

Mason, R.G. and Raff, A.D., 1961. Magnetic survey off the west coast of North America, 32°N 42°N latitude. *Bull. Geol. Soc. Am.*, 72: 1259–1266.

Maxwell, A.E., Von Herzen, R.P., Hsü, K.J., Andrews, J.E., Saito, T., Percival, S.F., Milow, E.D. and Boyce, R.E., 1970. Deep-sea drilling in the South Atlantic. *Science*, 168:1047–1059.

McElhinny, M.W., 1973. *Palaeomagnetism and Plate Tectonics*. Cambridge University Press, Cambridge, 358 pp.

McElhinny, M.W. and Evans, M.E., 1968. An investigation of the strength of the geomagnetic field in the early Precambrian. *Phys. Earth Planet. Interiors*, 1:485–497.

McKenzie, D.P. and Parker, R.L., 1967. The North Pacific: an example of tectonics on a sphere. *Nature*, 216:1276–1280.

McKenzie, D.P. and Sclater, J.G., 1971. The evolution of the Indian Ocean since the Late Cretaceous. *Geophys. J. R. Astron. Soc.*, 25:437–528.

Meidav, T., 1960. Nomograms to speed up seismic refraction computations. *Geophysics*, 25:1035–1053.

Menard, H.W., 1969. The deep-ocean floor. *Sci. Am.*, 221:126–142.

Mescherikov, Y.A., 1963. Secular movements of the earth's crust. In: *Recent Movements of the Earth's Crust, 1*. Publishing House for U.S.S.R. Academy of Sciences, Moscow, pp.7–24.

Meyerhoff, A.A., 1970. Continental drift, I. Implications of paleomagnetic studies, meteorology, physical oceanography and climatology. *J. Geol.*, 78:1–51.

Meyerhoff, A.A. and Teichert, C., 1971. Continental drift, III. Late Palaeozoic glacial centres and Devonian-Eocene coal distribution. *J. Geol.*, 79:285-321.

Montfrans, H., 1971. *Palaeomagnetic Dating in the North Sea Basin.* Doctoral Thesis, University of Amsterdam, Amsterdam, 113 pp.

Mooney, H.M. and Bleifuss, R., 1953. Magnetic susceptibility measurements in Minnesota, 2. Analysis of field results. *Geophysics*, 18:383-393.

Moorbath, S., O'Nions, R.K., Pankhurst, R.J., Gale, N.H. and McGregor, V.R., 1972. Further rubidium-strontium age determinations on the very early Precambrian rocks of the Godthaab District, West Greenland. *Nat. Phys. Sci.*, 240:78-82.

Morelli, C., 1970. Special issue on basement mapping by magnetics. *Boll. Geofis.*, 12, 182 pp.

Morgan, W.J., 1968. Rises, trenches, great faults and crustal blocks. *J. Geophys. Res.*, 73:1959-1982.

Morgan, W.J., 1971. Convection plumes in the lower mantle. *Nature*, 230:42-43.

Morley, L.W. and Larochelle, A., 1964. Palaeomagnetism as a means of dating geological events. *R. Soc. Can., Spec. Publ.*, 8:512-521.

Mörner, N.A., Lanser, J.P. and Hospers, J., 1971. Late Weichselian palaeomagnetic reversal. *Nat. Phys. Sci.*, 234: 173-174.

Moxham, R.M., 1963. Natural radioactivity in Washington County, Maryland. *Geophysics*, 28:262-272.

Mueller, S., 1968. Low-velocity layers within the earth's crust and mantle. *Proc. 10th Gen. Assembly European Seismol. Comm., Leningrad.*

Mueller, S., 1970. Geophysical aspects of Graben formation in continental rift systems. *Int. Upper Mantle Project, Sci. Rep.*, No.27, pp.27-37.

Mueller, S. (Editor), 1974. The structure of the Earth's Crust. *Int. Upper Mantle Project, Sci. Rep.*, No. 39, 391 pp.

Mueller, S. and Landisman, M., 1966. Seismic studies of the earth's crust in continents: evidence for a low velocity zone in the upper part of the lithosphere. *Geophys. J. R. Astron. Soc.*, 10:525-538.

Mueller, S., Peterschmitt, E., Fuchs, K. and Ansorge, J., 1969. Crustal structure beneath the Rhinegraben from seismic refraction and reflection measurements. *Tectonophysics*, 8:529-542.

Musgrave, A.W. (Editor), 1967. *Seismic Refraction Prospecting.* Society of Exploration Geophysicists, Tulsa, Okla., 604 pp.

Nafe, J.E. and Drake, C.L., 1963. Physical properties of marine sediments. In: M.N. Hill (Editor), *The Sea, 3.* Interscience, New York, N.Y., pp.794-815.

Nagata, T., 1961. *Rock Magnetism.* Maruzen, Tokyo, 350 pp.

Nagata, T. and Ozima, M., 1967. Paleomagnetism. In: S. Matsushita and

W.H. Campbell (Editors), *Physics of Geomagnetic Phenomena*, *1*. Academic Press, New York, N.Y., pp.103-180.

Nairn, A.E.M., 1964. *Problems in Palaeoclimatology*. Interscience, New York, N.Y., 705 pp.

Naudy, H., 1970. Une méthode d'analyse fine des profiles aéromagnetiques. *Geophys. Prospect.*, 18:56-63.

Néel, L., 1955. Some theoretical aspects of rock magnetism. *Adv. Phys.*, 4:191-243.

Nettleton, L.L., 1939. Determination of density for reduction of gravimeter observations. *Geophysics*, 4:176-183.

Nettleton, L.L., 1940. *Geophysical Prospecting for Oil*. McGraw-Hill, New York, N.Y., 444 pp.

Nettleton, L.L., 1942. Gravity and magnetic calculations. *Geophysics*, 7:293-310.

Nettleton, L.L., 1954. Regionals, residuals and structures. *Geophysics*, 19:1-22.

Nettleton, L.L., 1971. *Elementary Gravity and Magnetics for Geologists and Seismologists*. Society of Exploration Geophysicists, Tulsa, Okla., 121 pp.

Ninkovich, D., Opdyke, N.D., Heezen, B.C. and Foster, J.H., 1966. Palaeomagnetic stratigraphy, rates of deposition and tephrachronology in North Pacific deep-sea sediments. *Earth Planet. Sci. Lett.*, 1:476-492.

Niskanen, E., 1939. On the upheaval of land in Fennoscandia. *Isostat. Inst., Helsinki, Publ.*, No.6, 30 pp.

Norris, D.K. and Black, R.F., 1961. Application of palaeomagnetism to thrust mechanics. *Nature*, 192:933-935.

Oliver, J. and Isacks, B., 1967. Deep earthquake zones, anomalous structures in the upper mantle, and the lithosphere. *J. Geophys. Res.*, 72:4259-4275.

Opdyke, N.D., 1968. The palaeomagnetism of oceanic cores. In: R.A. Phinney (Editor), *The History of the Earth's Crust*. Princeton University Press, Princeton, N.J., pp.61-72.

Orellana, E. and Mooney, H.M., 1966. *Master Tables and Curves for Vertical Electrical Sounding over Layered Structures*. Interscience, Madrid, 193 pp.

Ozima, M., Ozima, M. and Kaneoka, I., 1968. Potassium-argon ages and magnetic properties of some dredged submarine basalts and their geophysical implications. *J. Geophys. Res.*, 73:711-723.

Pakiser, L.C., 1963. Structure of the crust and upper mantle in the western United States. *J. Geophys. Res.* 68:5747-5756.

Pakiser, L.C. and Zietz, I., 1965. Transcontinental and upper mantle structure. *Rev. Geophys.*, 3:505-520.

401

Parasnis, D.S., 1961. *Magnetism.* Hutchinson, London, 128 pp.
Parasnis, D.S., 1971. Physical property guide for rocks and minerals. *A.B.E.M. (Stockholm) Geophys. Mem.,* No.4/71, 12 pp.
Parasnis, D.S., 1972. *Principles of Applied Geophysics.* Chapman and Hall, London, 213 pp.
Parasnis, D.S., 1973. *Mining Geophysics.* Elsevier, Amsterdam, 2nd revised and up-dated ed., 395 pp.
Pavoni, N., 1964. Aktive horizontal Verschiebungszonen der Erdkruste. *Bull. Ver. Schweiz. Petrol. Geol. Ing.,* 31:54 78.
Pavoni, N., 1966. Recent horizontal movements of the earth's crust as related to Cenozoic tectonics. *Ann. Acad. Sci., Fennicae, Ser. AIII,* 90:317-324.
Pavoni, N., 1969. Zonen lateraler horizontaler Verschiebung in der Erdkruste und daraus ableitbare Aussagen zur globalen Tektonic. *Geol. Rundsch.,* 59:56-77.
Pekeris, C.L., 1940. Direct method of interpretation in resistivity prospecting. *Geophysics,* 5:31-42.
Pemberton, R.H., 1970. Airborne radiometric surveying for mineral deposits. *Geol. Survey Can., Econ. Geol. Rep.,* No.26, pp.416-424.
Peters, L.J., 1949. The direct approach to magnetic interpretation and its practical application. *Geophysics,* 14:290-319.
Platou, S.W., 1970. The Svaneke granite complex and the gneisses on East Bornholm. *Bull. Geol. Soc., Denmark,* 20:93-133.
Poley, J.H. and Van Steveninck, J., 1970. Geothermal prospecting. *Geophys. Prospect.,* 18:666-700.
Poster, C.K., 1973. Ultrasonic velocities in rocks from the Troodos Massif, Cyprus. *Nat. Phys. Sci.,* 243:2-3.
Pratt, J.H., 1855. On the attraction of the Himalayas Mountains and of the elevated regions beyond upon the plumb-line in India. *Phil. Trans. R. Soc., Lond., Ser. B.,* 145:53-100.

Radhakrishnamurthy, C. and Likhite, S.D., 1966. Initial susceptibility and constricted Rayleigh loops of basalts. *Curr. Sci.,* 35:534.
Raitt, R.W., 1963. The crustal rocks. In: M.N. Hill (Editor), *The Sea, 3.* Interscience, New York, N.Y., pp.85-102.
Ramberg, I. and Lind, G., 1968. Gravity measurements on the Paarup Salt dome. *Bull. Geol. Soc., Denmark,* 18:221-240.
Reford, M.S. and Summer, J.S., 1964. Aeromagnetics: a review. *Geophysics,* 29:482-516.
Reich, H., 1932. Die Bedeutung der finnischen Schweremessungen für die angewandte Geophysick. *Gerlands Beitr. Geophys.* 2(2):1-13.
Richter, C.F., 1935. An instrumental earthquake magnitude scale. *Bull. Seismol. Soc. Am.,* 25:1-32.

402

Riddihough, R.P., 1972. Regional magnetic anomalies and geology in Fennoscandia: a discussion. *Can. J. Earth Sci.*, 9:219–232.

Rikitake, T., 1952. Electrical conductivity and temperature in the earth. *Bull. Earthquake Res. Inst., Tokyo Univ.*, 30: 13–24.

Ringwood, A.E., 1969. Composition and evolution of the upper mantle. In: P.J. Hart (Editor), *The Earth's Crust and Upper Mantle (Geophysical Monograh 13)*. American Geophysical Union, Washington, D.C., pp.1–17.

Ringwood, A.E. and Green, D.H., 1966. Petrological nature of the stable continental crust. In: J.S. Steinhart and T.J. Smith (Editors), *The Earth Beneath the Continents (Geophysical Monograph 10.)* American Geophysical Union, Washington, D.C., pp.611–619.

Roche, A. and Wohlenberg, J., 1970. Magnetic measurements in Alsace, Baden and Pfalz. *Int. Upper Mantle Project, Sci. Rep.*, No.27, pp.224–228.

Roy, A., 1961. On some properties of residuals and derivatives. *J. Geophys. Res.*, 66:543–548.

Roy, A., 1966. The method of continuation in mining geophysical interpretation. *Geoexploration*, 4:65–84.

Rudman, A.J. and Blakely, R.F., 1965. A geophysical study of a basement anomaly in Indiana. *Geophysics*, 30: 740–761.

Runcorn, S.K., 1956. Palaeomagnetic comparisons between Europe and North America. *Proc. Geol. Assoc. Can.*, 8:77–85.

Runcorn, S.K., 1962. Palaeomagnetic evidence for continental drift. In: S.K. Runcorn (Editor), *Continental Drift*. Academic Press, London, and New York, N.Y., pp.1–39.

Rutten, M.G. and Wensink, H., 1960. Palaeomagnetic dating, glaciations and the chronology of the Plio-Pleistocene in Iceland. *Int. Geol. Congr., Sess. 21, Part IV*, pp.62–70.

Rybach, L., 1973. *Wärme Produktionsbestimmungen an Gesteinen der Schweizer Alps*. Inst. für Geophysik E.T.H., Zürich, 43 pp.

Sass, J.H., 1964. Heat flow values from eastern Australia. *J. Geophys. Res.*, 69:3889–3993.

Sass, J.H., 1971. The earth's heat and internal temperatures. In: I.G. Gass, P.J. Smith and R.C.L. Wilson (Editors), *Understanding the Earth*. Artemis Press, Sussex, pp.81–87.

Saxov, S., 1956. Some gravity measurements in Thy, Mors and Vendsyssel. *Geodaet. Inst. Skrifter, Ser. 3*, 25, 46 pp.

Scheller, E., 1970. *Geophysikalische Untersuchungen zum Problem des Taminer Bergsturzes*. Dissertation Nr.4560 E.T.H., Zürich, 91 pp.

Scholander, P.F., Dansgaard, W., Nutt, D.C., DeVries, H., Coachman, L.K. and Hemmingsen, E., 1961. Radiocarbon age and oxygen-18 content of Greenland icebergs. *Medd. Grønland*, 165(1), 25 pp.

Schønemann, A., 1972. *Magnetic Investigations of Some Diabase Dykes on Bornholm.* Graduate Thesis, University of Copenhagen, 86 pp. (in Danish).

Scrutton, R.A. and Roberts, D.G., 1971. Structure of Rockall Plateau and Trough, northeast Atlantic. In: F.M. Delanev (Editor), *The Geology of the East Atlantic Continental Margin, 2. Europe.* Institute of Geological Sciences, London, Reprint No. 70/14, 170 pp.

Sharma, P.V., 1966. Rapid computation of magnetic anomalids and demagnetization effects caused by bodies of arbitrary shape. *Pure Appl. Geophys.*, 65: 89-109.

Sharma, P.V., 1967. Graphical evaluation of magnetic and gravity attraction of three-dimensional bodies. *Geophys. Prospect.*, 15:167 173.

Sharma, P.V., 1968. Choice of configuration for measurement of magnetic moment of a rock specimen with a fluxgate unit. *Geoexploration*, 6: 101-108.

Sharma, P.V., 1970. Geophysical evidence for a buried volcanic mount in the Skagerrak. *Bull. Geol. Soc., Denmark*, 19:368 377.

Sharma, P.V., 1971. Tables of solid angles and potential field functions for geophysical applications. *Inst. Geophys. Univ. of Copenhagen, Publ.*, No.3, 105 pp.

Sharma, P.V., 1973. Seismic velocity and sediment thickness investigations by refraction soundings in Nûgssuaq, West Greenland. *Geol. Survey Greenland, Rep.*, No.54, 22 pp.

Sharma, P.V., 1975. Magnetic properties of some Tertiary basalts from West Greenland. *Pure Appl. Geophys.*, in press.

Sholpo, L.E., 1967. Regularities and methods of study of the magnetic viscosity of rocks. *Izv. Akad. Nauk, U.S.S.R.*, 6:99-116.

Slotnick, M.M., 1959. *Lessons in Seismic Computation.* Society of Exploration Geophysicists, Tulsa, Okla., 278 pp.

Small, J.B., 1963. Interim Report on vertical crustal movement in the United States. *Paper presented at the 13th Gen. Assembly, I.U.G.G., Berkeley, Calif.*, 1963.

Smith, A.G., 1971. Continental drift. In: I.G. Gass, P.J. Smith and R.C.L. Wilson (Editors), *Understanding the Earth.* Artemis Press, Sussex, pp.213-231.

Smith, A.G. and Hallam, A., 1970. The fit of the southern continents. *Nature*, 225:139-144.

Smith, P.J., 1970. The intensity of the ancient geomagnetic field: a summary of conclusions. In: S.K. Runcorn (Editor), *Palaeogeophysics.* Academic Press, London, pp.79-90.

Smith, R.A., 1959. Some depth formulae for local magnetic and gravity anomalies. *Geophys. Prospect.*, 7:55 63.

Smith, R.A., 1960. Some formulae for interpreting local gravity anomalies. *Geophys. Prospect.*, 8:607–613.

Smithson, S. B., 1963. Granite studies: I. A gravity investigation of two Precambrian granites in South Norway. *Norg. Geol. Undersök.*, 214B: 54–140.

Sørensen, H., Hansen, J. and Bondesen, E., 1969. Preliminary account of the geology of the Kvanefjeld area of the Ilímaussaq intrusion, South Greenland. *Geol. Survey Greenland, Rep.*, No. 18, 40 pp.

Sorgenfrei, T., 1969. Geological perspectives in the North Sea area. *Bull. Geol. Soc., Denmark*, 19:160–196.

Sorgenfrei, T., 1971. On the granite problem and the similarity of salt and granite structures. *Geol. Föreningen Stockholm Förhandlingar*, 93(2):371–435.

Srivastava, S. P., Hyndman, R. D. and Cochrane, N. A., 1973. Magnetic and telluric measurements in Atlantic Canada. *Geol. Survey Can. Paper*, No. 71–23:359–370.

Stacey, F. D., 1969. *Physics of the Earth*. John Wiley and Sons, New York, N.Y., 324 pp.

Stacey, F. D. and Banerjee S. K., 1974. *The Physical Principles of Rock Magnetism*, Elsevier, Amsterdam, 244 pp.

Steenland, N. C., 1965. Oil fields and aeromagnetic anomalies. *Geophysics*, 30: 706–739.

Stoner, E. C., 1945. Demagnetization factors for ellipsoids. *Phil. Mag.*, 36:803–821.

Storetvedt, K. M., 1968. On remagnetization problems in palaeomagnetism. *Earth Planet., Sci. Lett.*, 4: 107–112.

Strangway, D. W., 1966. Rock magnetism and geologic correlation. In: *Mining Geophysics, 1*. Society of Exploration Geophysicists, Tulsa, Okla., pp.54–66.

Strangway, D.W., 1970. *History of the Earth's Magnetic Field*. McGraw-Hill, New York, N.Y., 168 pp.

Summers, G. C. and Broding, R.A., 1952. Continuous velocity logging. *Geophysics*, 17:598–614.

Sykes, L. R., 1967. Mechanism of earthquakes and nature of faulting on mid-oceanic ridges. *J. Geophys. Res.*, 72:2131–2153.

Sykes, L. R., 1968. Seismological evidence for transform faults, sea-floor spreading and continental drift. In: R. A. Phinny (Editor), *History of the Earth's Crust*. Princeton University Press, Princeton, N.J., pp. 120–150.

Symons, D. T. A., 1967. The magnetic and petrologic properties of a basalt column. *Geophys. J. R. Astron. Soc.*, 12: 473–490.

Tagg, G. F., 1964. *Earth Resistance*. George Newnes, London, 258 pp.

Talwani, M., 1965. Computation with the help of a digital computer of magnetic anomalies caused by bodies of arbitrary shape. *Geophysics.*, 30:797–817.

Talwani, M. and Ewing, M., 1960. Rapid computation of gravitational attraction of three-dimensional bodies of arbitrary shape. *Geophysics*, 25:203–225.

Talwani, M., Worzel, J. L. and Ewing, M., 1957. Gravity anomalies and crustal structure of the Bahamas. *Trans. 2nd Carib. Geol. Conf.*, pp. 151–161.

Talwani, M., Le Pichon, X. and Ewing, M., 1965. Crustal structure of the mid-ocean ridges, 2. Computed models from gravity and seismic refraction data. *J. Geophys, Res.*, 70:341–352.

Takeuchi, H., Uyeda, S. and Kanamori, H., 1970. *Debate about the Earth.* Freeman, Cooper and Company, San Francisco, Calif, rev. ed., 281 pp.

Tarling, D. H., 1967. The palaeomagnetic properties of some Tertiary lavas from East Greenland. *Earth Planet. Sci. Lett.*, 3:81–88.

Tarling, D.H., 1971. *Principles and Applications of Palaeomagnetism.* Chapman and Hall, London, 164 pp.

Thellier, E. and Thellier, O., 1959. Sur l'intensité du champ magnétique terrestre dans le passe historique et géologique. *Ann. Geophys.*, 15: 285–376.

Uyeda, S., 1962. Thermoremanent magnetism and reverse thermoremanent magnetism. In: T. Nagata (Editor), *Proceedings Benedium Symposium on Earth Magnetism.* University of Pittsburgh Press, Pittsburgh, Ill., pp. 87–106.

Vacquier, V., 1972. *Geomagnetism in Marine Geology.* Elsevier, Amsterdam, 185 pp.

Vacquier, V. and Affleck, J., 1941. Computation of the depth of the bottom of the earth's magnetic crust. *Trans. Am. Geophys. Union*, pp.446–450.

Vacquier, V. and Uyeda, S., 1967. Palaeomagnetism of nine seamounts in the western Pacific and of three volcanos in Japan. *Bull. Earthquake Res. Inst., Tokyo Univ.*, 45:815–848.

Vacquier, V., Steenland, N.C., Henderson, R.G. and Zietz, I., 1951. Interpretation of aeromagnetic maps. *Geol. Soc. Am., Mem. No.47*, 151 pp.

Van Dam, J.C., 1967. Mathematical denotation of standard graphs for resistivity prospecting in view of their calculation by means of a digital computer. *Geophys. Prospect.*, 15:57–60.

Van Dam, J.C. and Meulenkamp, J.J., 1967. Some results of the geo-electrical resistivity method in ground water investigations in the Netherlands. *Geophys. Prospect.*, 15:92–115.

Van der Voo, R., 1969. Palaeomagnetic evidence for the rotation of the

Iberian Peninsula. *Tectonophysics*, 7:5-56.

Van Nostrand, R.G. and Cook, K.L., 1966. Interpretation of resistivity data. *U.S. Geol. Survey. Prof. Paper*, 499, 310 pp.

Vening Meinesz, F.A., 1929. *Theory and Practice of Pendulum Observations at Sea*. J. Waltman, Delft.

Vening Meinesz, F.A., 1948. Gravity expeditions at sea, 1923-1938. *Publ. Nederlands Geod. Comm.*, 4:1-233.

Vening Meinesz, F.A., 1964. *The Earth's Crust and Mantle*. Elsevier, Amsterdam, 124 pp.

Verhoogen, J., 1959. The origin of thermoremanent magnetization. *J. Geophys. Res.*, 64:2441-2449.

Verhoogen, J., 1960. Temperatures within the earth. *Am. Sci.*, 48:134-159.

Verhoogen, J., 1962. Oxidation of iron-titanium oxides in igneous rocks. *J. Geol.*, 70:168-181.

Vine, F.J., 1968. Magnetic anomalies associated with mid-ocean ridges. In: R.A. Phinney (Editor), *The History of the Earth's Crust*. Princeton University Press, Princeton, N.J., pp.73-89.

Vine, F.J., 1971. Sea-floor spreading. In: I.G. Gass, P.J. Smith and R.C.L. Wilson (Editors), *Understanding the Earth*. Artemis Press, Sussex, pp.233-250.

Vine, F.J. and Hess, H., 1970. Sea-floor spreading. In: A.E. Maxwell (Editor), *The Sea*, 4. John Wiley and Sons, New York, N.Y., pp.587-622.

Vine, F.J. and Matthews, D.H., 1963. Magnetic anomalies over oceanic ridges. *Nature*, 199:947-949.

Vozoff, K. and Swift, Jr., C.M., 1968. Magnetotelluric measurements in the North German Basin. *Geophys. Prospect.*, 16:454-473.

Ward, M.A., 1963. On detecting changes in the earth's radius. *Geophys. J.R. Astron. Soc.*, 8:217-225.

Watkins, N.D., 1968. Comments on the interpretation of linear magnetic anomalies. *Pure Appl. Geophys.*, 69:179-192.

Watkins, N.D., 1969. Non-dipole behaviour during an Upper Miocene geomagnetic polarity transition in Oregon. *Geophys. J.R. Astron. Soc.*, 17:121-149.

Wegener, A., 1928. *The Origin of Continents and Oceans* (English transl., 1966, by J. Biram). Dover, New York, N.Y., 246 pp.

Werner, S., 1945. Determination of the magnetic susceptibility of ores and rocks from Swedish iron ore deposits. *Sver. Geol. Undersök. Årsbok*, 39:1-79.

Wilson, J.T., 1963. A possible origin of the Hawaiian Islands. *Can. J. Phys.*, 41:863-870.

Wilson, J.T., 1965. A new class of faults and their bearing upon continental drift. *Nature*, 207:343–347.

Wilson, J.T., 1966. Did the Atlantic close and reopen? *Nature*, 211:676–681.

Wilson, J.T., 1973. Mantle plumes and plate motions. *Tectonophysics*, 19:149–164.

Wilson, R.L., 1966. Palaeomagnetism and rock magnetism. *Earth-Sci. Rev.*, 1:175 212.

Wilson, R.L., 1970. Permanent aspects of the earth's non-dipole magnetic field over Upper Tertiary times. *Geophys. J. R. Astron. Soc.*, 19:417–437.

Wollard, G.P., 1969. Tectonic activity in North America as indicated by earthquakes. In: P.J. Hart (Editor), *The Earth's Crust and Upper Mantle (Geophys. Monograph 13)*. American Geophysical Union, Washington, D.C., pp.125–133.

Wollard, G.P., 1970. Evaluation of the isostatic mechanism and role of mineralogic transformations from seismic and gravity data. *Phys. Earth Planet. Interiors*, 3:484–498.

Wollin, G., Ericson, D.B. and Ryan W.B.F., 1971. Magnetism of the earth and climatic changes. *Earth Planet. Sci. Lett.*, 12:175–183.

Woods, J.P., 1956. The composition of reflections. *Geophysics*, 21:261–276.

Worzel, J.L., 1965a. *Pendulum Gravity Measurements at Sea, 1936 1959*. John Wiley and Sons, New York, N.Y., 422 pp.

Worzel, J.L., 1965b. Deep structure of coastal margins and mid-oceanic ridges. In: W.F. Whittard and R. Bradshaw (Editors), *Submarine Geology and Geophysics*. Butterworths, London, pp.335–359.

Wyllie, P.J., 1971. *The Dynamic Earth*. John Wiley and Sons, New York, N.Y. 416 pp.

York, D. and Farquhar, R.M., 1972. *The Earth's Age and Geochronology*. Pergamon Press, London, 178 pp.

Zietz, I. and Andreasen, G.E., 1967. Remanent magnetization and aeromagnetic interpretation. In: *Mining Geophysics, II*. Society of Exploration Geophysicists, Tulsa, Okla., pp.569–590.

Zohdy, A.A.R., 1965. The auxiliary point method of electrical sounding interpretation and its relationship to the Dar Zarrouk parameters. *Geophysics*, 30:644 660.

Index

410

418